Worlds beyond

WORLDS

IAN RIDPATH

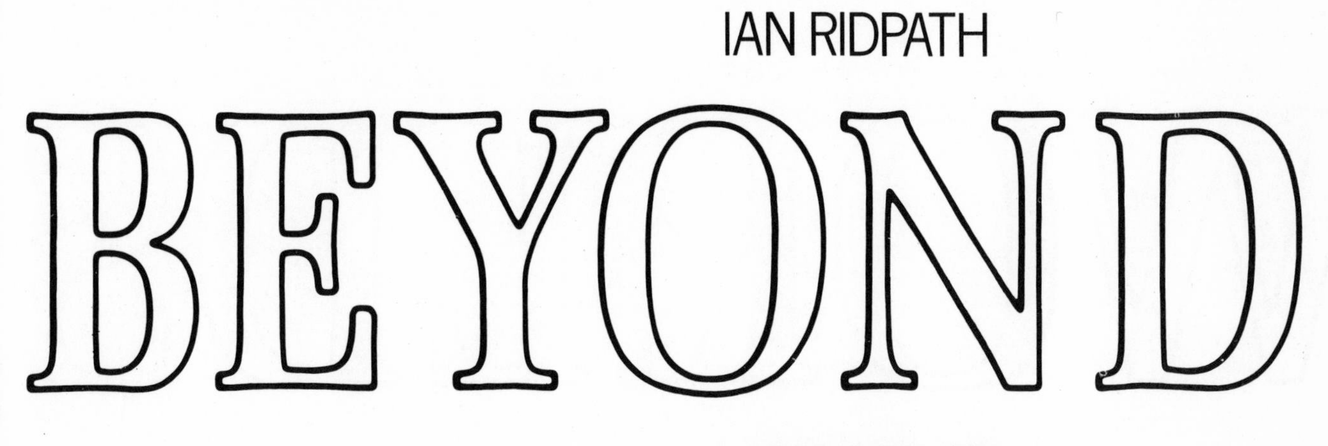

BEYOND

A REPORT ON
THE SEARCH FOR LIFE
IN SPACE

HARPER & ROW, PUBLISHERS

New York, Hagerstown, San Francisco, London

Quotations in Chapter 9 from *UFOs–A Scientific Debate,* © 1972 by Cornell University Press, are reproduced by kind permission of the publishers.

This book was originally published by Wildwood House Limited, 29 King Street, London, in 1975.

ISBN: 0-06-013568-9

LIBRARY OF CONGRESS CATALOG CARD NUMBER: 75-30344

76 77 78 79 80 10 9 8 7 6 5 4 3 2

FOR ANN,
who really started it all

CONTENTS

BACKGROUNDER

This book grew out of a feature I wrote for the magazine section of the *Observer* newspaper. My thanks go to John Silverlight for backing that first article, and to Oliver Caldecott of Wildwood House for allowing me to expand the story into book form.

But there would be no story to write without the efforts of those scientists who have been patiently unravelling the steps by which nature pieced us all together. In particular, I thank those scientists who took time out to talk with me, in every case providing new knowledge and clarification of technical points. They appear like characters throughout this book—for, despite popular opinion, science is as much about people as the knowledge that they uncover.

Particularly I should like to thank the astronomers of Queen Mary College, London, who helped socially as well as factually, and encouraged me to use their library for research.

The final draft of this book was prepared during quiet days in Worcestershire, thanks to the hospitality of Ilene and Richard, widow and son of Donald Suddaby whose writings in science fiction have pleased generations of readers.

Those who were particularly helpful in collecting information or photographs and are not otherwise mentioned or credited in the book: Geoffrey Falworth, A. Bruce Whitehead (JPL), John Billingham (NASA), George Sands (NASA), Christine Gravell (Macdonald & Co.), Paul Sutherland (*Hermes*), Enid Lake (Royal Astronomical Society), Wallace Oref (NRAO), Ms S.M. Hayward (British Museum, Natural History), L.G. Marsh (Hulton Educational Publications).

Gratitude also to Chris Priest, who helped, and Pat Mumford, likewise. And to my family, who allowed it to happen.

Ian Ridpath
February 1975

My writing is clear as mud, but mud
settles and clear streams run on and
disappear, perhaps that is the reason
but really there is no reason except
that the earth is round and that no
one knows the limits of the universe
that is the whole thing about men and
women that is interesting

Gertrude Stein
Everybody's Autobiography

INTRODUCTORY

A CRY IN THE NIGHT

Our civilization is awaiting a telephone call with a difference: placed by people whom we do not know, whom we will not recognize, whom we may not even understand. The call will be coming ultra-long distance.

A call from space.

According to one American astronomer, whole volumes of the Encyclopedia Galactica may be winging their way towards us, packed with all the scientific knowledge that an advanced and benevolent society might consider important—the secrets of longevity and peaceful co-existence, the answer to malignant diseases, perhaps new sources of power that dwarf even the atom's explosive stockpile. Interspersed with these electrifying revelations could be instructions for learning Galactic, the language of interstellar communication—an open invitation to us to answer the call.

But would the possible dangers of answering a call from another civilization outweigh the likely benefits? The Czech-British astronomer Zdenek Kopal has visualized our confrontation with extraterrestrial beings intelligent enough to have discovered our existence: 'We might find ourselves in their test tubes or other contraptions set up to investigate us as we do guinea pigs.'

Though some believe that the possible gain in information would be a major benefit of interstellar communication, Kopal is in no doubt that the motives of other civilizations are not to be trusted. 'If the space phone rings,' he pleads, 'for God's sake let us not answer it.' These are the philosophical problems with which astronomers currently grapple. The debate is not held in public at the moment, but it is a discussion that increasingly enters the public domain.

We should not feel snugly isolated from the hostile Universe on our small planet Earth, assuming that it is a notional problem, imagining that no one can know we are here. They can.

Since the invention of radio, we have been broadcasting a clarion-call to space. Our voices have already ascended into the sky, and vanished into the blackness of space like a cry in the night.

Who—or what—will respond to our call?

Nor should anyone think that a return message would go unheeded. Many are already listening. Daily, the radio telescopes of the world run the chance of stumbling across a cosmic call-sign. Specific attempts are made to eavesdrop on other stars. And others search nearby space for evidence of alien space probes.

These scientific events are reflected in a groundswell of public opinion about life in space. A Gallup survey in North America released on November 28th, 1973, and reported in the *New York Times*, reveals the extent to which people are ready primed to accept the sound of a voice from space: 46 per cent of those polled believe that there is intelligent life on other planets.

This poll was taken at a time when a strange comet fever gripped the world, as the newly discovered Comet Kohoutek ghosted towards the Sun like a cosmic ectoplasm. In that same month, the growing space consciousness of the young generation was highlighted when, after a special Comet Kohoutek presentation, the New York Hayden Planetarium threw open its doors at midnight to a party welcoming the British space-rock group, Hawkwind.

Under the approving eye of the planetarium's director, respected astronomer Dr Kenneth Franklin, the planetarium's co-ordinator of public affairs, Richard Hoagland, commented, 'Hawkwind represents the forefront of a new cultural interface for society—an interest in astronomy. The interest may be an emotional one, and we want to season that emotionalism with a little scientific knowledge.'

In the hot San Francisco summer of 1972, thousands of people coiled round the Palace of Fine Arts Theatre and snarled up traffic for miles around in vain attempts to gain admission to the first of a series of lectures on Cosmic Evolution, organized jointly by NASA, the City College of San Francisco, and the Astronomical Society of the Pacific.

Second-night speaker, cosmologist Geoffrey Burbidge, reported that fighting among those unable to enter the packed hall for his lecture had resulted in broken windows. Thereafter, further publicity was suspended and the lectures were given twice-nightly. This success is particularly significant in an area that a NASA spokesman described as 'the heart of enemy territory'—the breeding-ground for anti-science movements that flourish in reaction to man's technological advance.

As the Pioneer 10 spacecraft hoves into the vicinity of Jupiter, its chemical apparatus sniffing for a clue to life, a private citizen has tapped out a warning in a letter to the British magazine *New Scientist*: 'Were the human race not so arrogant, those members of it who possess superior intelligence and knowledge should already have considered that they might, just possibly, not be the peak of biological evolution.'

In November 1974 a businessman in Leeds, England, bet £25 with Ladbroke's that alien beings will land on Earth by 1975. He got odds of 1000 to 1.

And in the psychological backwash of Watergate, the greatest U.F.O. flap since Kenneth Arnold sighted the first 'flying saucer' in 1947 gripped the United States. In that same American Gallup poll, taken after the post-Watergate U.F.O. flap had peaked, 51 per cent of those sampled believed that U.F.Os are real, and not the result of imagination or hallucination; 11 per cent claimed to have seen one themselves.

There was a strong link between those who believed in the reality of U.F.Os and those who speculated about life elsewhere. Clearly, U.F.Os and life in outer space are linked in the public mind. And a poll among scientists about the possibility of life elsewhere would bring similar results.

The U.S. National Academy of Sciences has said that contact with other civilizations in space is 'a natural event in the history of mankind that will perhaps occur in the lifetime of many of us.'

Such scientific beliefs rest not on U.F.O. observations—which many frankly reject—but on a statistical inevitability wrought by the sheer size of our Universe. In the simplest terms, they take the view that we are nothing special. If intelligent life can arise here, then given enough time and chance it can happen over again anywhere else. The insignificance of Earth in the Universe has been drummed into scientists, who in recent history have been forced to remove first the Earth, then the Sun, and finally our Galaxy, from the hub of their thinking.

Astronomer-biologist Carl Sagan says, 'The idea of extraterrestrial life is an idea whose time has come.'

In a historic meeting in 1961, a group of scientists discussed how they might calculate the probability of life's existence elsewhere in the Universe. The group, meeting at America's National Radio Astronomy Observatory, was presented with a formula filled with unknown quantities to which they tried to put numbers. They came out with a result which showed that it is entirely possible that another intelligent civilization is nurtured by at least one of the stars visible in the compass of our eyes.

Surveying the views of scientists as they stood in the mid 1960s, prize-winning science journalist Walter Sullivan of the *New York Times* concluded tersely, 'We are not alone!'

Astronomer-philosopher Fred Hoyle has spoken of the vast interstellar telephone traffic that must be going on between advanced civilizations. Carl Sagan says, 'We may be like the inhabitants of the valleys of New Guinea who may communicate by runner or drum, but who are ignorant of the vast international radio and cable traffic passing over, around, and through them.' Zdenek Kopal has written wonderingly, 'Are such messages being beamed to us while you read these pages?' And Arthur C. Clarke, doyen of science-fiction writers, pondering the carrying power of our transmissions since the megawatt radars of World War II, has been moved to speculate, 'We have been making such a din that the neighbours can hardly have overlooked us, and I sometimes wonder when they will start banging on the walls.'

Faced with a message from an alien civilization, we could be as psychologically numbed as men from a Stone Age culture thrown headlong into consumer society. The reaction is known as culture shock.

American sociologist Alvin Toffler defines this as the effect that immersion in a strange culture has on the unprepared visitor. It disorients those who travel from nation to nation on our own Earth, 'when', he says, 'the familiar psychological clues that help an individual to function in society are suddenly withdrawn and replaced by new ones that are strange or incomprehensible.'*

How much greater would the culture shock be if brought about by beings from another planet?

Toffler has also diagnosed the disease of future shock among those numbed by civilization's current headlong advance, which scientifically and technologically has already propelled us into areas where society's response lags far behind. In laboratories throughout the world, scientists challenge our traditional beliefs by their search to create life. We seem too future-shocked already even to contemplate the outcome of a trans-stellar meeting of minds.

Outside the realms of science fiction, what work has been done holds out a prospect of radical change. A study-report commissioned by NASA on the eve of man's first flights into space pointed out:

> Anthropological files contain many examples of societies which have disintegrated when they have had to associate with previously unfamiliar societies espousing different ideas and different life ways; others that survived such an experience usually did so by paying the price of changes in values and attitudes and behaviour.

* *Future Shock* (Bodley Head, London, 1970).

We also need to resolve the problem of *who* should talk to the stars. At a Soviet-American conference on Communication with Extraterrestrial Intelligence held in 1971 in Byurakan the delegates concluded, 'It seems to us appropriate that the search for extraterrestrial intelligence should be made by representatives of the whole of mankind.'

The scientists will have to decide these problems for themselves, for they will get no political lead—nor would they necessarily follow it.

Science is not the only discipline for which the prospect of life in space involves much heart searching. Religions that contend that God made man in His own image would be severely shaken should we find another creative, intelligent race made in a different image. And clerics still debate whether each civilization would need its own Saviour.

Several established religions have faced, and answered, these questions with the affirmation that other life in space is to be expected. At a symposium in 1972 at Boston University theist Krister Stendahl declares, 'It is great when God's world gets a little bigger and I get a bigger view.'

Since deciding that life on other worlds is consistent with Marxist ideology, the Soviet Union has become prominent in advocating that it probably exists.

But now, new religions blossom from the space consciousness that provides the roots of present-day philosophy. Flying-saucer cults believe that we are being watched by cosmic masters, and some claim to receive messages from them. Today's manifestations are not of the blessed Virgin, but of flying-saucer spacemen.

Yet such views suffer from the same fallacy that flaws the philosophy of religion: 'The U.F.O. seers are incredibly geocentric in their conceits,' says geneticist Joshua Lederberg of Stanford University.

Astronomer Thomas Gold even suggests light-heartedly that life on Earth could have arisen from the debris dumped by interstellar travellers who briefly visited Earth. Thus we become no better than the maggots of space.

While cynics joke that the Earth and its faulty life-forms were an early, and abandoned, experiment by an immature God, for others the uneasy suspicion remains that Men have created god in *their* own image.

One man who clings to rational belief is astronomer Sir Fred Hoyle, champion of cosmological ideas that deny a creation for the Universe. Hoyle calls the urge of scientists to study the physical world a religion in itself. He compares the motives of today's scientists, especially astronomers searching for the clues to decode the Universe, with the motives of the builders of Stonehenge—more truly religious than 'the crude ritualistic survivals from the Stone Age' that pass for organized religion today.

What is it that nourishes the current space consciousness? Is it merely dissatisfaction with the answers offered by our preachers? Why do so many of us now seek for signs among the stars?

'In a Universe whose size is beyond human imagining, where our world floats like a dust mote in the void of night, men have grown inconceivably lonely,' wrote anthropologist Loren Eiseley in *The Immense Journey*,* a book that appeared in 1957, ten years after the term 'flying saucer' was coined.

* Victor Gollancz, London, 1958.

Men were frightened by the night long before religion arrived to give them temporary comfort. Men's spirits have remained unchanged from the time that they were apes and reached for the Moon like Moon-Watcher in Arthur C. Clarke's *2001*. Dreams of flying did not begin with the Wright brothers, any more than did nightmares of falling begin with the first tall buildings. Stories of Moon flight begin in literature soon after the advent of the printed book. Stories of sky-men go back to the earliest records, such as in the Old Testament.

Our need to explore the Universe is deeply grained in our nature. Space efforts now focus mass attention on the sky, altering the conscious outlook of modern man.

The major turning point can be dated precisely—Christmas Eve in 1968, when the Moon-orbiting crew of Apollo 8 became the first men to see the Earth rise over the landscape of another body in space. From the vicinity of the Moon, astronaut James Lovell, peering through binoculars, commented in awe, 'The Earth looks pretty small from here.' And he also wondered aloud, 'What would a traveller from another planet think of Earth. And where would he land?'

Lovell summed up his feelings shortly before the return to Earth: 'The vast loneliness of the Moon is awe inspiring, and it makes you realize just what you have back there on Earth. The Earth from here is a grand oasis in the big vastness of space.'

And Frank Borman coined his famous phrase, 'the Good Earth'.

In league with this, the realization that the Earth is only an ordinary planet in space has led to a greater recognition of our own frailties.

In 1973 Dr Ernst Stuhlinger, a former German rocket scientist at Peenemünde, and now associate director of science at the Marshall Space Flight Center in Huntsville, Alabama, answered a letter from Sister Mary Jucunda, a nun renowned for her work among the starving peoples of Africa. His reply to her question about the benefits of space exploration contained this passage:

> Although our space programme seems to lead us away from our Earth and out towards the Moon, the Sun, the planets, and the stars, I believe that none of these celestial objects will find as much attention and study by space scientists as our Earth.
>
> It will become a better Earth, not only because of all the new technological and scientific knowledge which we will apply to the betterment of life, but also because we are developing a far deeper appreciation of our Earth, of life, and of man.

Referring to the classic Apollo 8 photograph of the rising Earth, Dr Stuhlinger continued:

> Of all the many wonderful results of the space programme so far, this picture may be the most important one.
>
> It opened our eyes to the fact that our Earth is a beautiful and most precious island in an unlimited void, and that there is no other place for us to live but the thin surface layer of our planet, bordered by the bleak nothingness of space.
>
> Never before did so many people recognize how limited our Earth really is, and how perilous it would be to tamper with its ecological balance.
>
> Ever since this picture was published, voices have become louder and louder warning of the grave problems that confront man in our times: pollution, hunger, poverty, urban living, food production, water control, overpopulation.
>
> It is certainly not by coincidence that we begin to see that tremendous task waiting for us at a time when the young space age has provided us the first good look at our own planet.

In the years following Apollo 8, another eight manned spacecraft made the run to the Moon from the Good Earth, six of them pausing to drop and pick up men on its surface.

Expanding his own space consciousness—and possibly trying out a future means of space communication—an astronaut conducted his own private E.S.P. experiment with friends on Earth during one such mission. Writing afterwards of the results obtained, the astronaut, Apollo 14's Captain Edgar Mitchell, said, 'In my judgement the experiment warrants continued inquiry which I shall hope to pursue as opportunity permits.'

Then, in 1973, came the largest spacecraft yet launched—the Skylab space station. After 48 days in orbit, the Sun rising and setting once every 93 minutes, the members of the third Skylab crew reported at the start of 1974 their 'almost spiritual' feelings: 'My attitude towards life is going to change,' said Air Force colonel, William Pogue.

The mission's scientist, Dr Edward Gibson, said, 'I think this mission is going to increase my awareness of what else is going on besides what I'm doing.' Gibson said that he had begun to 'speculate a little more' about life elsewhere in space. 'Being up here and being able to see the stars and look back at the Earth and see your own Sun as a star makes you much more conscious of that,' he said. 'You realize the Universe is quite big and just the number of possible combinations that you can have out there which can create life enters your mind and makes it seem much more likely.'

But to see the scientific reasons for that likelihood, we have to take a closer look at the fabric of our night-sky surroundings.

COUNTING OUR NEIGHBOURS

Look up into the heavens on a cloudless night. If you live under averagely clear skies, you will be able to spot about a thousand stars above you. They twinkle as they shine, these jewels in the black velvet of the heavens, and closer study shows that they vary not only in brightness, but also in colour.

What are the stars, and how far away are they?

Only one star is visible during the daytime: the Sun. The Sun is our own star, and it is the one we know most about. It appears different from other stars because, relatively speaking, it is exceptionally close.

Approximately 93 million miles separate us from the Sun. A typical star seen by naked eye in the night sky is around a million times more distant. But, peering into the abyss of infinity, the greatest telescopes of the world span voids where distance has a new meaning. The mile is such a puny measure in the vastness of space, that astronomers have turned to another yardstick. They chose the fastest speed known to man—the speed of light. Distances in space are measured by the time it takes a light-beam to travel them. Light travels so swiftly that it can circle the Earth in the snap of a finger—186,282 miles in a second. It spans the gulf from Earth to Moon in $1\frac{1}{4}$ seconds. But it takes over 8 minutes to reach us from the Sun. Even riding on a morning sunbeam, we could find time to eat breakfast between leaving the Sun and arriving at the Earth.

In the jargon of space, the Sun is more than 8 light minutes away from us. Yet the nearest other star is $4\frac{1}{4}$ light years away. Such distances are typical of the isolation of stars.

Nor does a radio signal flashed into the night sky hold out a hope of quicker communication with a star. Radio waves, like X rays, ultraviolet light, and infra-red radiation, are all part of the same spectrum; *the* spectrum whose visible portion we see as light and which Isaac Newton chopped into its constituent colours in his famous experiments with prisms. And every part of that spectrum snakes through space at the same speed as light. Light's limiting speed is the barrier to instantaneous knowledge of the Universe.

The lights we see in tonight's sky have been on their way to us for many years. If we could extinguish each star at the flick of a switch, we should still spend the rest of our lives watching them blink out one by one. Yet we can be fairly sure that none of those stars has ceased to shine. This confident assertion comes from our knowledge of the inner fires that power all stars.

The dating of rocks, based on the declining power of their radioactive atoms that decay with age, shows that the Earth has been around for thousands of millions of years. Meteorites, pieces of primordial matter that plunge to Earth, indicate that solid lumps of matter have existed in the Sun's neighbourhood for 4,650 million years. The Sun must have been shining for at least that length of time.

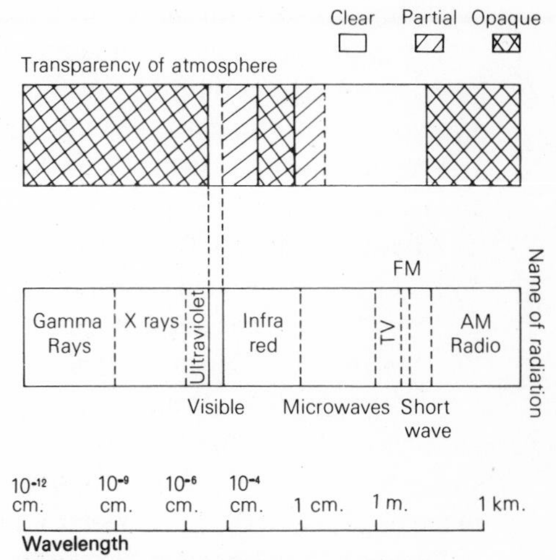

The spectrum, from short wavelengths to long. The visible region lies in the centre, at the wavelengths which our Sun emits most strongly. The Earth's atmosphere lets only a part of the total spectrum through to us on the surface of the Earth.

What power can fuel a fire for so many aeons? Brilliant minds were teased for centuries by this celestial riddle. Processes common on Earth could not provide the answer. If the Sun glowed like a lump of coal, it would be a charred ember after a few thousand years. This theory may have been acceptable before geological science was founded, in the days when Archbishop James Ussher of Armagh in 1650 added up the ages of the saints and pronounced the Earth's origin to have been in 4004 B.C. But towards the end of last century, the founders of geology, scratching the surface of our globe for clues to its past, came to realize that the story of the Earth must be reckoned in far longer chapters.

All seemed to be satisfactorily explained when astronomers calculated that a slow contraction of the Sun would release sufficient energy to feed its current brilliance for 10 to 100 million years. Such a span seemed fit to encompass the entire history of God's Universe, not just the recent history of Earth and its inhabitants recounted by the Scriptures.

Yet we now know this explanation to be woefully inadequate.

It is only in the past few decades that we have plotted the true length of the Sun's past—and have discovered the source of its mysterious energy. This revelation required the birth of an entirely new physics, dealing in processes not even hinted at a century ago. For nuclear energy is the source of the Sun's power.

Strange rays from a compound of uranium, fogging the photographic plate of the French physicist Henri Becquerel in 1896, gave a vital clue to the power that lay hidden in the atom. Becquerel, by his chance observation, had discovered radioactivity.

Patient probing of the atom, during the first years of this century, revealed that atoms have a core, surrounded by electrons. The atomic core, or nucleus, is unstable in some atoms. The energy of radioactive decay is caused by changes in the nucleus. Occasionally, the core can be prised apart by adding energy. The atom then splits into two lighter atoms, releasing a far greater burst of energy. This is the *fission* process, which powered the atomic bomb and which is controlled at a sleepy pace to raise turbine-spinning steam-heat in electricity-generating stations.

Yet the process can also work the other way. Crushing light atomic cores together builds a heavier atom. This is nuclear fusion. It, too, releases power—far more, even, than fission.

The growing science of nuclear physics during the 1930s gave glimpses of understanding of the Sun's energy processes. By the end of World War II, the centre of the Sun had, figuratively, been laid bare: hydrogen, the main stuff of the Universe, was being turned into helium by nuclear fusion, the same process that causes the destructive flash of a hydrogen bomb. But a star, being a massive thing, is held together by its own gravity, and does not fly apart—luckily for us.

Apart from the natural radioactivity in rocks which heats the Earth's interior, fusion seems to be the only atomic process that is important on the cosmic scale. The scalding heat of the Sun is something that scientists struggle to contain in their experimental fusion reactors, which may one day supplant the relatively puny, but more easily tamed, fission reactors.

Fusion works because the nucleus of a helium atom weighs slightly less than the four hydrogen nuclei that fuse to form it. This extra mass is turned into energy during the fusion reaction, in accordance with Einstein's famous $E = mc^2$ equation. Around this equation, mystique has grown. Yet its meaning is simple to read: 'E' is the energy produced; 'm' is the mass of hydrogen that goes to produce it; 'c' is the speed of light. In one second, the nuclear furnace at the centre of the Sun converts about 600 million tons of hydrogen into helium. Another 4 million tons of hydrogen are annihilated; they are turned into energy to keep the Sun aglow.

Despite the fact that the Sun is losing mass at such a prodigious rate, it is so big that it can last without wasting away for thousands of millions of years. And it is those thousands of millions of years that have seen the emergence of a new and remarkable creature on Earth, with the audacity of intellect to unlock the secrets of its own origin.

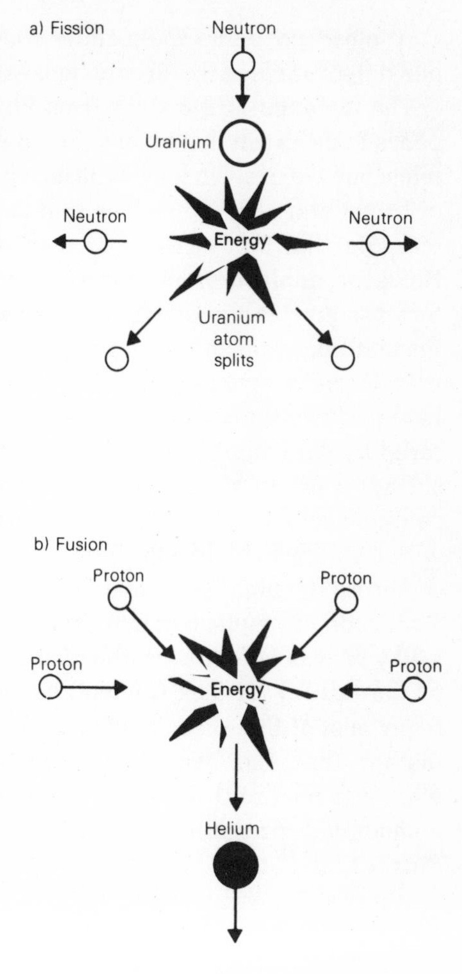

Atomic energy processes. It is the continual process of fusion that powers the stars. In the Sun, the cores of hydrogen atoms, called protons, crush together to make helium.

The Sun is an average star in every way. Some stars are much bigger, some are much brighter, and some are much hotter. Equally, others are smaller, fainter and cooler. The colour of a star is a direct guide to its temperature. Stars burn red-hot, white-hot, or blue-hot: the blue being the hottest. The Sun's yellow-white surface is typical, but a close look at some of those thousand pinpoints in the night sky reveals blue stars, red stars, and all shades in between.

Without the Sun we would not exist: its light provides the energy to make things grow, and its heat makes life comfortable. Stars produce the power that fuels the engines of life.

The Sun also affects us in other subtle ways: our eyes are adjusted so that they detect just those wavelengths that the Sun emits most strongly. Civilizations bred under the spotlight of a different star would develop eyes that see the world in a different way.

The Sun's diameter of 865,000 miles dwarfs even the largest planet. It would take over 100 Earths marshalled side by side to cross this distance.

Can anyone truly visualize a body of this size? In thinking about the Universe, we should realize that the Sun, which we take for granted, is already beyond our comprehension. Other statistics about the Sun only confirm how puny our imaginations are. We should need to cram a third of a million Earths into a scale pan to outweigh the Sun. And a jar with the volume of the Sun would hold 1.3 million Earth-sized bodies.

This is the essential difference between a star and a planet: a planet is too small for nuclear reactions to be sparked off in it, whereas it is the enormous pressures and temperatures generated at the heart of a star that tinder its fire. Planets shine in the night sky simply because they reflect the light of the Sun. They appear bright because of their proximity to us compared with the stars.

The planets are held in orbit around the Sun by the invisible thread of gravity. Various bits of debris left over from the formation of the planets, such as the rubble-filled comets, also swirl round the Sun. Together, these orbiting objects make up what is termed the Solar System.

The Earth is the third planet from the Sun. Nearest of all to the Sun is tiny Mercury, then the Earth-sized Venus. Beyond the Earth and its attendant Moon lie rocky Mars, then the giant, gaseous Jupiter, the ringed Saturn, and the smoky-blue dimness of Uranus and Neptune. An interloper at the Solar System's edge, Pluto, slowly weaves in and out across the orbit of Neptune.

We can easily pick out the planets from their starry background, because in their regular orbiting of the Sun they move noticeably over several nights.

In contrast, the stars are so distant that they appear to us as relatively fixed. Any planets that they might have would be too faint for us to see.

A telescope shows the planets of our solar system as disks, and we can detect surface features on most of them.

The distance of the stars from Earth means that none of them appears to us as anything more than a point of light, no matter what size telescope we use. On long-exposure photographs, the burnt-out images of bright stars make blotches that may be mistaken for the shape of a disk. But the Sun is the only star whose face we can see directly. However, analysis of observations using a technique called 'speckle interferometry' is starting to reveal disk details on certain stars, such as the giant Betelgeuse.

In its path through the Earth's turbulent atmosphere, starlight is bent and twisted so that it twinkles. The light from some stars is shattered by the Earth's atmosphere as if having passed through a prism, so that a bright star near the horizon will often flash and glint with rainbow colours. Yet if we were closer to such a star it would shine with a steady light, like the Sun.

Across the night sky trails the Milky Way, represented by the artist Tintoretto as a splash of milk from the breast of the goddess Juno. The Milky Way is in fact a swathe of stars, all loosely bound by the gravity of the Galaxy. It took Galileo's telescope, in 1609, to show that the foggy mist of the Milky Way is made of countless myriads of stars, so distant that the eye alone cannot see them individually. 'Upon whatever part of it the telescope is directed, a vast crowd of stars is immediately presented to view,' he wrote in *Sidereus Nuncius* (The Starry Messenger).

We now know that when we look at the Milky Way we are viewing the rim of our Galaxy from the inside. Our Galaxy is a pinwheel-shaped spiral of stars, only the nearest of which we see in the night sky. Astronomers estimate that it contains perhaps a million million stars, smeared across an area so big that light takes a hundred thousand years to cross it.

Avenues of stars curl out from the Galaxy's bulging heart. By counting the number of stars visible in various directions, the astronomer William Herschel reported in 1784 that our own Galaxy was a flat slab of stars 'in which the Sun is placed, though perhaps not in the very centre of its thickness.'

Although this notion of our Sun being nearly central persisted until early this century, more detailed work has shown that the Sun, with its attendant family of planets, lies about three-fifths of the way to one edge of the Galaxy. Only a small part of the Galaxy's spread is visible to astronomers with optical telescopes, but radio astronomers trace the radiation from cool hydrogen gas that may one day form into stars. The gas follows the vortex shape that we see for other galaxies.

One such galaxy is the famous Andromeda spiral, visible as a distant smudge in the evening skies of a northern winter. This is the farthest that the human eye can penetrate the reaches of space: two million light years. The light we see from the Andromeda galaxy started out while our ancestors were fighting duller apes on the scorched plains of East Africa.

In a culinary analogy, British sky watcher Patrick Moore describes a side-on view of our Galaxy as resembling two fried eggs clapped back to back. The thickness of the whites represents the galactic plane, seen in profile at night as the Milky Way.

On this scale the Solar System, reduced to an invisible atom, sees only the surrounding atoms as the individual stars in its sky. The star-stuffed yolk is too distant, and too obscured by dust, to be clearly seen.

It is among the closest stars, distributed fairly randomly around us

Millions of stars cluster together in a dense part of the Milky Way. Among these, might one have a planet on which life has arisen? (*Hale Observatories*)

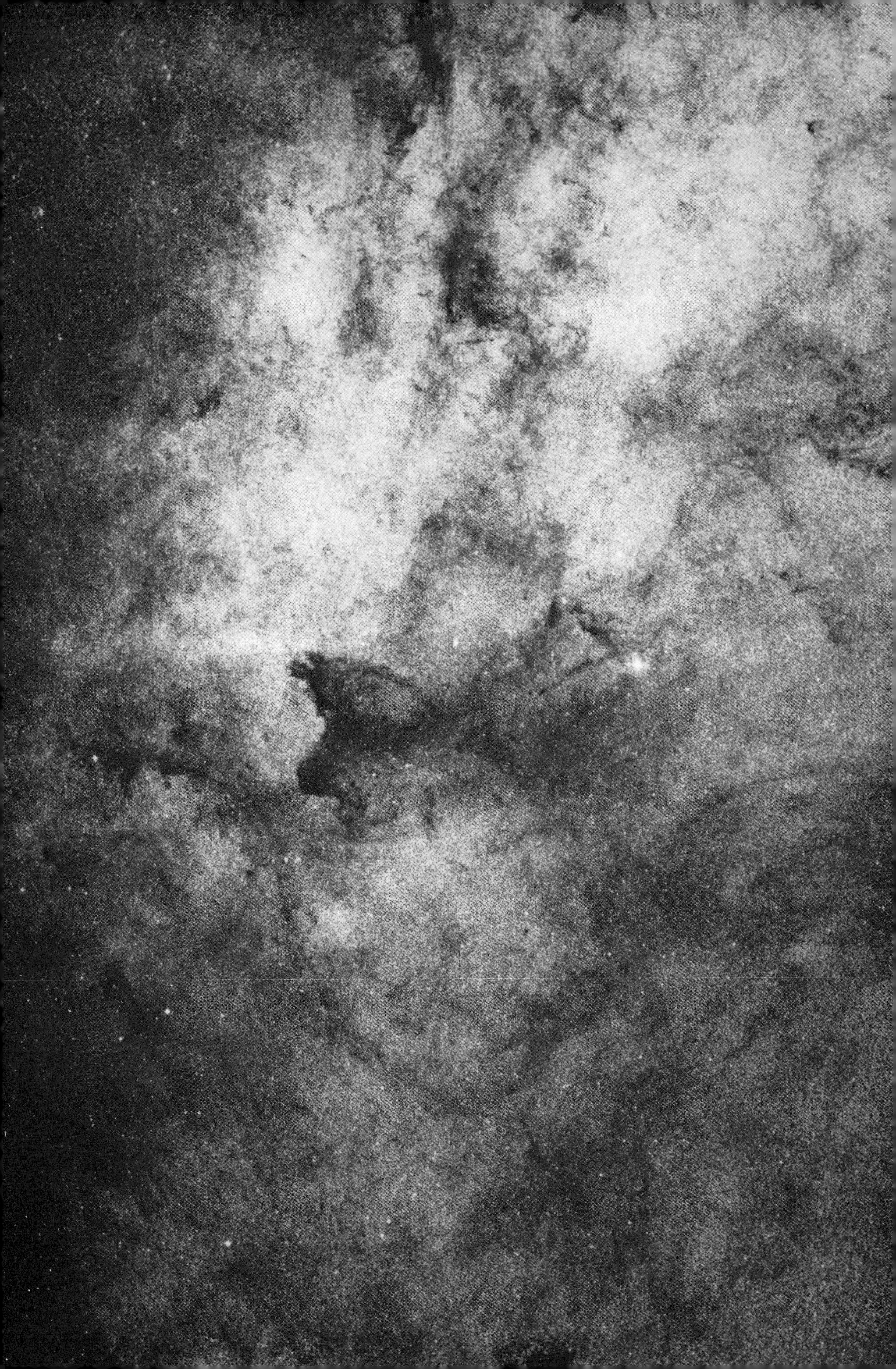

The spiral galaxy M 101 in Ursa Major. The Milky Way would appear similar to this if seen from outside. (*Lick Observatory*).

in our area of the Galaxy, that we draw the pictures called constellations. These represent mythological figures, such as Orion and Hercules, imposed somewhat arbitrarily on the unyielding star patterns by romantic-minded, or superstitious men of ancient times. Most of our popular star names come from ancient writings of Mediterranean and Middle Eastern peoples.

In the wake of Galileo, as more telescopes began to pierce the night, astronomers discovered the cloudy nebulae in our Galaxy. Dark 'holes', such as the famous Coal Sack of the southern hemisphere, seemed to have been drilled out of the Milky Way. These dark patches are caused where clouds of dust and gas blanket the light from stars behind. Elsewhere, the nebulae glow brightly: such dust-and-gas clouds, astronomers have since found, are the assembly-yards of stars.

Some nebulae presented more of a puzzle. They had a regular structure, later found in some to resemble a spiral. Astronomers spent many centuries debating the nature of these.

Sir William Herschel once remarked that he had discovered more than 1,500 'universes' (we would say galaxies). In 1755 the philosopher Immanuel Kant in his book *Universal Natural History and Theory of the Heavens* speculated that '. . . these elliptical objects are just universes—in other words, Milky Ways.' As recently as 1920, two American astronomers met in public debate on the question of whether the spiral nebulae were island universes—galaxies in their own right—or whether they were condensed star-clouds in our own Galaxy, like fragments of the Milky Way. Within the following few years it was finally proved that our Galaxy of stars was but one of many, spread at random throughout the Universe.

The farther that mankind looks into space, the more galaxies he sees. Optical and radio telescopes are grasping weak beams from galaxies that shone brightly before the Earth was born. Their radiation is reaching us after thousands of millions of years' travel through the Universe.

Galaxies vary in shape and size: some are coagulations of stars, like elliptical blobs, and others have spoke-like bars. But the flattened

A cluster of distant galaxies. Each faint smudge on this picture is a galaxy roughly the size of the Milky Way. Hard, round images are stars near to us. (*Hale Observatories*).

spiral of our own Galaxy seems fairly typical both in shape and size. Thus we see again that there is nothing special about our local environment in space. And we realize what a massive canvas we contemplate when assessing our own importance in the Universe.

To guess how many civilizations exist that we might contact in space, we must apply a formula that runs as follows. It is the formula that was presented by Frank Drake to a historic meeting of scientists in 1961, held at the National Radio Astronomy Observatory, Green Bank, West Virginia to discuss the possibility of extraterrestrial life. There, the scientists multiplied together seven different factors: rate of star formation at the time of our own Sun's birth; the fraction of those stars with planets; the percentage of such planets suitable for life; the number on which life actually does arise; the likelihood of intelligence among such life; the desire of that life-form to communicate; and, finally, the average longevity of civilizations, like our own, that satisfy all the above requirements.

The meeting at which these unknowns were first discussed was held under the auspices of the Space Science Board of the National Academy of Sciences. It was graced by the presence of some of the most prominent scientists in the search for man's origins. The scientists opened bottles of champagne during the proceedings when the announcement came that one of them, chemist Melvin Calvin, had been awarded a Nobel prize.

If we can accurately quantify each of the unknowns in the formula, we can from pure theory find the number of communicative civilizations in the Galaxy. The results of the calculation are exciting—and perplexing.

Star formation at the time of our Sun's birth is an important factor because the type of life, and its degree of advancement, could be radically different for stars that formed much before or after the Sun.

Stars are being formed all the time in the Galaxy, as far as astronomers can tell. They seem to come into existence from giant clouds of gas spread among the arms of our Galaxy. Study of the stars shows that the oldest ones formed about 10,000 million years ago. These tend to be confined to a halo around our Galaxy, and were formed at a time when the Galaxy itself was taking shape.

These oldest stars are made of only the simplest material — hydrogen and helium. A mixture of these two gases is believed to have been the primitive substance of the Universe.

But the power-giving nuclear reactions inside stars that turn hydrogen into helium can also progress to produce all the other types of atom that we know. Astronomers term all these other atoms the *heavy elements*.

The largest stars, in which the most extreme reactions occur, become unstable at the ends of their lives, and explode. In the blinding flash of their deaths, which we see as supernovae, all the chemical elements of the Universe are formed and scattered into space. They mix with the existing hydrogen and helium and can later join in the formation of new stars.

We are formed from the remains of these exploded stars. The atoms of the Earth and the atoms of our bodies were once formed in the nuclear cauldron of a supernova explosion.

Blow-up. The explosion of a star as a supernova scatters heavy elements into space — and leaves a pulsar at the centre. (*Hale Observatories*).

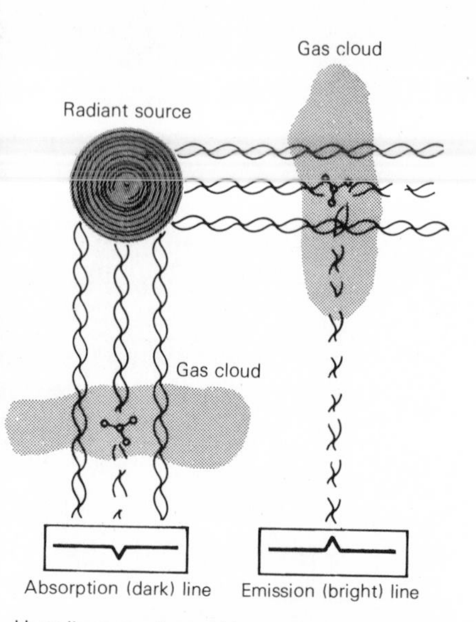

How lines are formed in a spectrum by absorption and emission of radiation by atoms in a gas cloud. The same effects work at radio wavelengths.

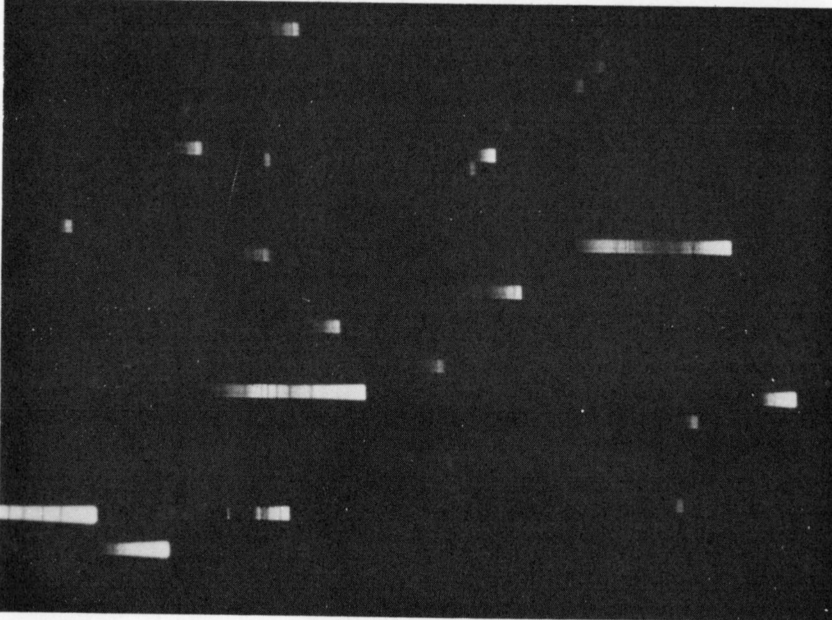

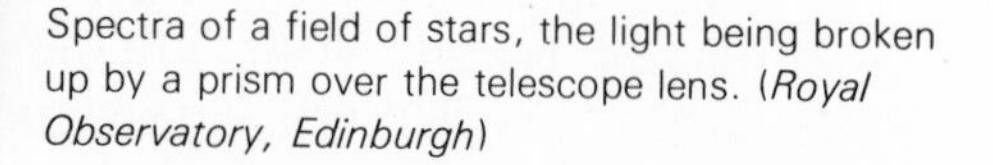

Spectra of a field of stars, the light being broken up by a prism over the telescope lens. (*Royal Observatory, Edinburgh*)

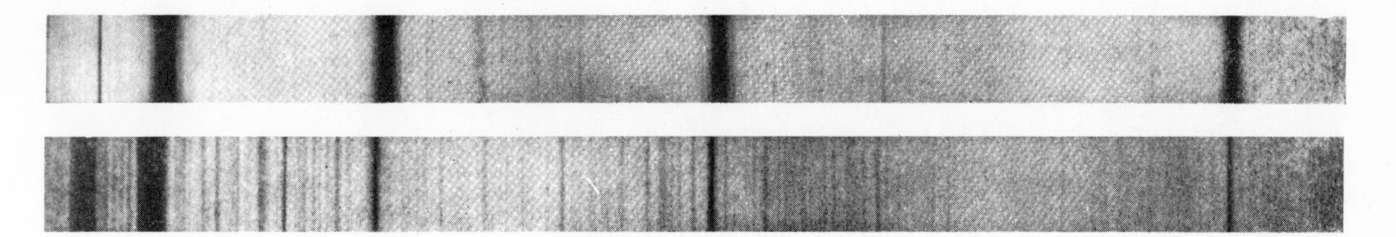

Absorption lines of hydrogen and other elements in the spectra of the bright stars Sirius and Procyon. The lines are caused by absorption of light in the cooler outer layers of the star.

The most recently formed stars have the highest proportion of heavy elements in them, because they are built from the wreckage of the many generations of stars that have gone before them. These elements give a kind of visual call-sign; that is, a distinctive pattern of thread-like lines across the rainbow spectrum of the star's light. Astronomers analyse starlight using a spectroscope—a sophisticated version of the simple prism that analyses light into its constituent colours. The spectroscope gives them a spectrum in which the atoms' call-signs are clear. Picking through the strands of starlight for the tell-tale marks of each generation, astronomers classify the stars of the Galaxy into populations of differing age. The first stars to be formed can have no planetary systems with living creatures, because the rocky material of planetary bodies and the chemical elements of life were not at that stage formed. Equally, the newest stars are too young for life to have evolved to the complex pitch of intelligence that life on Earth has developed. So we should restrict our interest to stars of similar age to the Sun—bearing in mind that our own civilization just one human generation ago could not communicate with the stars. And one human generation ahead—if we are not extinct—we shall almost certainly have originated techniques beyond our current understanding.

Unfortunately, this also applies to signalling. A suitably advanced civilization could be trying to reach us with techniques that are completely unknown (and undetectable) by us. They would no more use something as cumbersome as a radio signal than we would consider flashing the Sun's light by a mirror to gain their attention. Perhaps someone, somewhere, is just sitting and *thinking* very hard . . .

We estimate that about one star a year forms in our own Galaxy at present. At the Sun's birth, this rate was probably several times greater because there was more interstellar gas to be swept up into new stars.

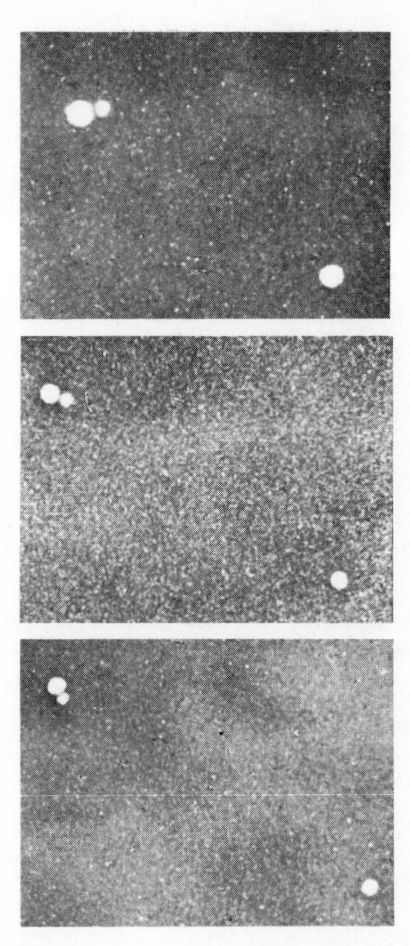

Rotation of the double star Kruger 60. (*Yerkes Observatory*)

About half the stars in the sky seem to form in twos, or larger groups. Life on a planet in a system with two or more stars, not necessarily of the same colour, presents an interesting possibility. But recent calculations have shown that planets orbiting double-star systems will either be forced to collide with one of the stars, or be thrown out of the system altogether.

So no more than half the stars of the Sun's age can have planetary systems. But current work on the origin of the planets shows that some form of entourage is an almost inevitable by-product of a star's growth. So those stars that remained single may well have a group of several planets round them. If our Solar System is as average as we think, then planets may be more numerous in the sky than stars.

Among such groups of planets, at least one planet will probably lie in the region around its star that is neither too warm nor too cold for life. There are several such planets in our own Solar System, though we have yet to discover whether life has arisen on more than our own.

Round the smallest and dimmest stars, the light and heat would be too feeble for a prodigious burst of life. The largest stars are too spendthrift with their energy, burning out—and exploding as supernovae—long before even the most primitive bacterium could gain a footing on a nearby planet. Some stars are too variable in light, and others flare dangerously and unpredictably, emitting beams of searing radiation into space. Although the environs of relatively few stars are fitted for life, astronomers, probing by radio into the dense regions where stars form, are finding the presence of rich life-creating materials. Carbon, nitrogen, oxygen, and the other key elements of our bodies cluster round the sites of forming stars.

The work of the past few years, reported in later chapters, supports in a concrete and spectacular way the bold assumption of the Green Bank group, who agreed that life of some sort was almost certain to arise, given long enough, on suitable planetary sites.

But *intelligent* life? That is another question. As we shall see below, not all scientists go along with the Green Bank participants, who decided that intelligence is as inevitable as the emergence of life itself.

Man is a notorious explorer and meddler, yet there are introspective tendencies in all of us. Would every intelligent species want to communicate—even if they did discover the technology enabling them to talk to the stars? We cannot even be confident that more than about one civilization in every ten intelligent ones would rise to our technological level. And how intelligent are *we*? Arthur C. Clarke, in one of his perceptive asides, has suggested that the size of our brains may be maladaptive. We are inquisitive enough to meddle, but not bright enough to foresee the outcome of our actions . . .

We shall not know the answer to this question for a few million years yet. At the moment, most of us would settle for another thirty.

After the Green Bank conference, Frank Drake is reported to have pinned on his office door a notice asking, 'Is there intelligent life on Earth?' This humbling query apart, it proved relatively easy to insert reasonable figures into the rest of the formula. From his own deductions, Drake later wrote, 'About one new intelligent civilization appears in the Milky Way a year.'

At the Green Bank conference on Intelligent Extraterrestrial Life, a final estimate of between 10,000 and 1,000 million communicative civilizations in our Galaxy was arrived at. The nearest cosmic telephone caller could then be several thousand light years away—or as close as 10 light years away. In the first case, a chat would be tedious in the extreme. If Pontius Pilate had asked the stars, 'What do I do about the Christians?' his question might be received any century now. Even communicating over a 10-light-year gap would rule out the need for an emergency answering service. But it would be far more use to us to know that we could exchange greetings in a lifetime.

Carl Sagan, foremost proponent of the idea of life on other worlds.

The normally optimistic astronomer-biologist Carl Sagan of New York State's Cornell University, discussing the problem in 1966 in a book called *Intelligent Life in the Universe**—a title that in itself would have been scientific heresy not many years before—could only settle on a compromise figure. He came to the answer of one civilization like our own (or in advance of it) for every one hundred thousand stars, giving a round-trip time of several centuries for a message between two such averagely spaced peoples.

A colleague of Drake's, the German radio astronomer Sebastian von Hoerner, took the discussion further in a paper published in the December 8th, 1961 issue of the American magazine *Science*. He considered the possibility that a series of technological civilizations might arise and become extinct on the same planet. 'We assume that a state

* Holden-Day, New York, 1966

of mind not too different from our own will have developed at many places, but will have only a limited longevity,' he wrote.

Another civilization might have a tough job in getting a toehold if its predecessors had used mineral and fuel reserves at the same prodigious rate as ourselves. But a gap of several hundred million years would serve to stock up the globe again with these natural consumables.

Von Hoerner's calculations gave a probable figure of one communicating civilization per three million stars, with a likely separation of a thousand or so light years. He pointed out that we should scan the whole sky for signals rather than concentrate on individual stars. He even drew from his calculations the probability that the civilization's development would be equivalent to our own level in A.D. 14,000. 'Our chance of learning from them might be considered the most important incentive for our search,' he wrote. Depressingly conservative though this one-in-three-million figure may sound, it still implies the existence of several hundred thousand civilizations like ourselves—or far in advance—throughout our Galaxy.

Since then, a whole spectrum of guesses have been made. They vary from the hyper-optimistic, such as the view of German-American rocket scientist Krafft Ehricke—that between 1 and 10 per cent of stars in our Galaxy have planets with intelligent communities, giving a massive total of between 1,000 million and 10,000 million in the Galaxy at present—down to a round zero.

Microbiologist Peter Sneath calculated that, out of a probable 12 million life-bearing planets, there might be but one broadcasting civilization in our Galaxy—which sounds suspiciously as though we are alone at present. But Sneath regarded his estimates as so variable that we could be the only broadcasting civilization in 1,000 galaxies—or there could be 1,000 like us in our own Galaxy.

Engineer J.G. Kreifeldt added a new dimension in a paper published in 1971 in the Solar System journal *Icarus*. He reformulated the problem, taking into account the time from the origin of life to the onset of the communicative phase. Kreifeldt's complex mathematical treatment can be used to produce computer models that predict how the number of communicative civilizations varies with time. But, in general terms, Kreifeldt's treatment reduces to the Drake and von Hoerner formulae.

At a Soviet-American conference on Communication with Extraterrestrial Intelligence (CETI) held at the Byurakan Astrophysical Observatory in Soviet Armenia in 1971, the Drake formula was discussed for two and a half days by representatives of various branches of science. In all, fifty-four scientists from six countries met at this conference, held in sight of Mount Ararat, the reputed resting place of Noah's ark. Marvin Minsky, a researcher into machine intelligence, epitomized the mood of the conference with his remark, 'The probability of exactly making man is very small. But there is not just one pathway to the equivalent of man.'

Carl Sagan summarized the discussion about some of the factors involved in calculating the existence of civilizations in the Galaxy. Oddly, the Byurakan results differed little from the estimates made at Green Bank a decade before. The average rate of star formation: 'About 10 per year.' The fraction of stars with planets: 'Perhaps a half, a third, a quarter, something like that.' The number of planets per solar system that are conducive to life: 'I would guess it is around one. In our solar system it is certainly several.'

There was a difference of opinion about the origin of life. Biologists

Francis Crick, a Nobel-prizewinner for his part in unravelling the double-helix structure of D.N.A. in cells, and Leslie Orgel, an avid researcher on the origins of life, concurred that it remains difficult to estimate reliably the origin of life. Perhaps it can happen anywhere; or perhaps it is rare. Other participants were optimistic that intelligence is highly likely in living systems. 'The selective advantage of intelligence is enormous,' commented Sagan.

A major stumbling-block in scientists' minds was the origin of technological civilization: are the use of fire and the development of language both prime requisites for technology, that relatively few intelligent civilizations might be able to master? Fire is repellent to many creatures, and only occurs in an oxidizing atmosphere. Sagan estimated from the discussions that technological life would arise on one in a hundred suitable planets, and inserted these estimates into the calculation. He came up with the answer that one such civilization is formed in the Galaxy about every ten years.

The lifetimes of such civilizations remained the stumbling-block of calculations on how many other technological groups might exist at this moment. 'If civilizations destroy themselves shortly after arriving there may be no one for us to talk to but ourselves,' pointed out Sagan.

With an echo of Alvin Toffler's *Future Shock*, physicist John Platt summed up the present human condition: 'We are in a transition period on a scale such as no [Earth] society has ever encountered. The result is that we may oscillate, we may destroy ourselves, or we may reach a high-level steady state.

'It will be a new form of society, totally different from anything that has ever existed in the world before, as radically different as a new species—if we survive.'

Canadian anthropologist Richard Lee recalled a phrase of the Soviet astrophysicist Josef Shklovsky, who spoke of a civilization that lives and dies in a day, like a butterfly. But Lee argued for the positive effect of interstellar contact on a civilization: 'If we receive a communication we want to stick around to find out what the reply is to our communication.'

Assuming that a small percentage of societies reach long-term stability, Sagan added the final term into the formula to come up with an overall result: 'There are a million technical civilizations in the Galaxy. This corresponds to roughly one out of every 100,000 stars.'

At the Byurakan conference, Sebastian von Hoerner gave a new estimation of the likely distance between currently communicative civilizations. He said: 'Some astronomical estimates show that probably about 2 per cent of all stars have a planet fulfilling all known conditions needed to develop life similar to ours.

'If we are average, then on half of these planets intelligence has developed earlier and farther, while the other half are barren or underdeveloped.'

The distance between intelligent neighbours would be about 14 light years, he calculated—but then made again the assumption that 'nothing lasts for ever.' A loss of interest in technology could silence a highly intelligent race, or they could be overtaken by one of the many crises of super-development: overpopulation, nuclear self-destruction, and what he calls 'genetic deterioration' caused by the elimination, by medicine, of natural selection effects.

Making 'a free guess' at a likely lifetime of 100,000 years (Sagan had guessed 10 million), von Hoerner found the average distance to our nearest chatty neighbour to be 600 light years.

Such a distance means that only far-advanced civilizations could ever hold two-way conversations, because only they last long enough. Therefore we should expect to contact someone well in advance of ourselves—or to find that many lines go dead, as each civilization wipes itself out.

We may therefore be best advised just to listen in, on the assumption that there are powerful beacon-signals intended to attract newcomers such as ourselves. 'If we really are average,' said von Hoerner, 'the first communications will have been discussed already 5 billion years ago.'

Although such estimates are widely variable, and may involve massive errors, Sagan defended the exercise in these words: 'We want to decide whether it is out of the question to search for extraterrestrial intelligence. We have the technological capacity to do it. The question is, shall we proceed?'

If the estimates cannot exclude the possibility of extraterrestrial life, Sagan predicted, 'I think it is very likely that we are going to proceed to try and find it.'

Sagan returned to the question with a paper published in *Icarus* in 1973, in which he suggested a possible strategy for finding life. He emphasized the growing feeling that communicative civilizations are likely to be the longest-lasting ones, and added:

Such societies will have discovered laws of nature and invented technologies whose applications will appear to us to be indistinguishable from magic.

There is a serious question about whether such societies are concerned with communicating with us, any more than we are concerned with communicating with our protozoan or bacterial forbears.

He referred to a classification of the technological stage of civilizations, originated by the Soviet radio astronomer Nikolai Kardashev. A so-called Type I civilization has the same sort of power output as our own Earth. A Type II civilization, of greater technological ability, can harness the same luminosity as a star. The most advanced development, a Type III civilization, can tap power equal to that of an entire galaxy.

'Should there exist even one Type II civilization in the local group of galaxies, there will be a realistic possibility of securing an enormous quantity of information,' Kardashev had written in the March-April 1964 issue of the Soviet *Astronomicheskii Zhurnal.* 'The same holds for the existence of even one Type III civilization in the observable Universe.'

Making the usual assumptions, Sagan calculated that the average Type I civilization, of similar capability to ourselves, might be too far away to detect with existing instruments. The chances of finding a nearby Type I civilization are slim because their lifetimes are so short. Perhaps only one in a hundred reaches the stability that allows it to progress to the technological level of a Type II or Type III civilization.

We may therefore be restricted to detecting the much stronger transmissions expected from Type II and Type III civilizations. 'If only a tiny fraction of such civilizations are interested in antique communications modes they will dominate the interstellar communications traffic now accessible on Earth,' Sagan believes. Thus we might be best off using existing technology to search for such advanced civilizations, rather than press on hopefully in search of someone similar to ourselves.

A continuing divergence of views about extraterrestrial life was demonstrated by two contributors to the 1972 San Francisco Cosmic Evolution series of lectures. Summarizing his own talk, communications engineer Dr Bernard Oliver of the Hewlett-Packard electronics company said, 'Since life on Earth seems to have originated and evolved from natural causes operating in the primitive environment, and since this environment would be commonly repeated in other planetary systems, it now appears that life is a common phenomenon in the Universe.' Drawing on his knowledge of the latest research in astronomy (which is reported in later chapters), Oliver stated, 'We now can believe that the Universe is teeming with life, some of it more advanced than life on Earth.'

However, another contributor to the same series, Professor Freeman Dyson of Princeton's Institute for Advanced Study, had this to say: 'I do not believe we yet know enough about stars, planets, life, and mind to give us a firm basis for deciding whether the presence of intelligence in the Universe is probable or improbable.' Stating his belief that chemists and biologists had made their optimistic predictions from 'inadequate evidence', Dyson added, 'I consider it just as likely that no intelligent species other than our own has ever existed.'

Dyson's failure to qualify intelligence as meaning technological intelligence would bring a strong rebuttal from marine scientists, who are now convinced that the intelligence of dolphins, and possibly whales, is at least as great as a child of kindergarten age. Dolphins may even be capable of intellectual feats beyond our own capacities. As yet, we've not been clever enough to find out.

Biologist Dr Peter Molton, of the University of Maryland's laboratory of chemical evolution, hit back by comparing cries of 'We are unique' with the old dogma 'We are the centre of the Universe.'

But Dyson did call for 'unprejudiced observation'—the only way, in his view, to resolve the problem.

In the following chapters we shall review the scientific evidence that shows why Dyson's view of the improbability of life is in the minority. And we shall conclude by looking at what some people believe to be evidence that we have already been discovered by other beings.

But, to put our own existence into perspective, we first examine the emergence of one intelligent species—and where it may be headed in the future.

THE RISE OF MEN

What happened, 20 million years ago, to set us on the road to being men?

Land movements in East Africa cracked open a valley, scientists say, isolating a group of apes from their fellows. Bounded by geological formations that they could not overcome, this group of apes became the breeding-stock of a new and super-intelligent terrestrial form of life, whose descendants were to spread over the globe—and, eventually, over the Solar System. Other groups of apes, isolated elsewhere by geographical conditions, evolved differently. The rubber-legged first steps of a human infant remind us of our common ancestry with the chimpanzees, which arose in Africa, west of the Rift Valley between the Sahara in the north and the Congo river to the south. In the jungles of central Africa the gorilla emerged; on the isolated island of Madagascar the lemur developed. And, swinging among the trees of South East Asian forests, the orang-utan evolved into its human caricature.

All these primates show how isolated groups of a similar stock can evolve along separate lines to suit their surroundings. Each species remains confined to its localized geographical region—all, that is, except the resourceful and peripatetic primate called man.

The theme of evolution stems from the ideas of Charles Darwin. Darwin's famous book *The Origin of Species*, published in 1859, first hinted at the connection between apes and men at a time when he could adduce virtually no supporting evidence for his theory. Those who championed his views, notably Thomas Henry Huxley, were sailing a small and fragile vessel in a very stormy sea. For all anyone knew at that time, Archbishop Ussher's date for the origin of the Earth could have been true.

It is difficult to reconstruct the atmosphere of society in those days. Perhaps the only present-day equivalent to the sensation that the idea of evolution caused would be the discovery of life on another planet.

Darwin's sequel to *The Origin of Species* was *The Descent of Man*, published in 1871, in which he presaged the modern view: 'It is probable that Africa was formerly inhabited by extinct apes closely allied to the gorilla and chimpanzee; and as these two species are now man's nearest allies, it is somewhat more probable that our early progenitors lived on the African continent than elsewhere.'

At that time, the sole known member of man's ancestral tree was Neanderthal man—although its significance was clear to only the most advanced thinkers of the day. Neanderthal man is but one of the development stages of man's forebears, and is now considered to have been off the main line of our descent. The Neanderthals are usually classified as an early form of *Homo sapiens*. Human beings of all races on Earth today are members of the same species. We are all part of the general grouping called *Homo*; the species name is *sapiens*, meaning wise. Neanderthal man, even though he lived around a hundred thousand years ago, was physically very similar to us. But to satisfy our desire for uniqueness, we usually introduce the distinction of a sub-species, which gives the cumbersome titles of *Homo sapiens neanderthalensis* for the Neanderthals, and *Homo sapiens sapiens* for ourselves. Only when we go back half a million years or more to the Peking and Java types of men do we come to a different species—*Homo erectus*: upright man.

The pigeon-holing of each species under precise names disguises the fact that evolution modifies one type into another.

Even in the subtle world of sub-species, men have remained mentally and physically unchanged for at least the past 10,000 years. It was then, in the New Stone Age—the Neolithic—that men turned from a nomadic way of life to become farmers. 'From Neolithic times onwards there were no more structural changes, only cultural ones,' says anthropologist David Pilbeam of Yale University, who has closely surveyed the whole history of man. The Neolithic revolution was the starting-point of our current development—and of our current problems. Agriculture allowed the creation of a surplus of food and hence of a surplus of people. As man put down his roots, cities grew, fed by the bulging granaries. Once the hunting life was left behind, men were on the way to creating a new way of life—the technology of contemporary civilization.

It has happened once on this planet. Perhaps, if man vanished, some other animal would find the evolutionary potential for it to happen again.

The grain of man was moulded in the monkeys that evolved between 40 and 35 million years ago. They were a wide-ranging and adaptable division of the primates whose descendants include modern monkeys, the chimps, gorillas, orang-utans—and the wise man himself, *Homo sapiens*.

Primates began to flourish around 70 million years ago, when the tyranny of the dinosaurs died away. The small, shy animals that had hidden from these terrible lizards rapidly became bolder in their forays. In the forests of North America, Europe, Asia and Africa, fruit-eating primates evolved their tree-scrambling mode of existence. They had larger brains, better vision, and were more dexterous than their forebears. Zoologists classify them today as the prosimians. They live on, little changed, in the incongruous forms of the indris, the graceful lemurs, the goggle-eyed galagos or bush babies, and the slender-fingered tarsiers.

In studying these animals, we, in effect, are digging through millions of years to see a living tableau of human origins. In the living prosimians we can recognize the crude outline of ourselves, like reflections in a distorted mirror. The multi-million-year gap has had relatively little effect on the living prosimians, for the rise of more skilful primates has nudged them into life-style niches where they remain supreme. Those in Africa and Asia became nocturnal, while those on Madagascar (a chip of the African continent that has drifted away) were isolated from the higher primates that arose on the mainland.

Even these relatively weak-brained primates, scientists now realize, form complex social groupings in which the young learn the lessons of life that frame their reactions to future events. Anthropologists conclude that, perhaps 50 million years ago, early primates had already evolved the first groupings that still shape the society of modern man. Says David Pilbeam, 'Human sociability and intelligence have origins rooted deep in our pre-human ancestry.'

It was from these roots that new shoots sprouted around 40 million years ago, signalling the emergence of a still more complex branch in the tree of life. These were the higher primates, including the anthropoids, whose fossil-remains now dot the shelves in museums of man's past. The evolutionary path to the monkeys had already been primed by the highly developed social state of the prosimians. In their footsteps, the higher primates with greater reasoning capacity and dexterity were able to displace the prosimians, who died out where the competition became too strong or retreated into life-styles or habitats where they could develop on their own.

These higher primates that impressed themselves upon the world had larger brains, and greater ability to understand, interpret and mould the world to their ends. The fossil record of this advancement of the brain is, regrettably, far from clear. Yet, by 20 million years ago, the apes were clearly set on their evolutionary path. They arose in an area of central and eastern Africa that the University of Oxford's professor of anatomy, Sir Wilfrid Le Gros Clark, described as 'an experimental breeding station'. From these apes fork the evolutionary tines that lead to man.

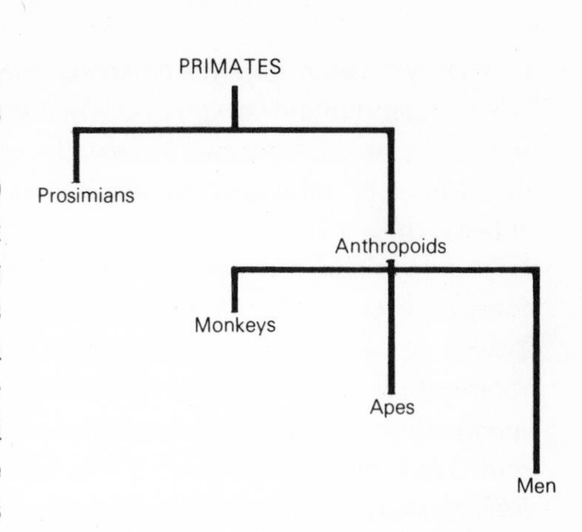

Schematic diagram of the primate family tree, showing different levels of evolutionary advance.

A new family of the anthropoids, called the Hominidae, were the harbingers of humanity. Their type was first seen on Earth perhaps 20 million years ago, when a group of apes evolved a new ground-feeding way of life at the forest edge, in clearings, and round lakes. These apes were medium in size, weighing around 40 pounds. In their new habitat, their skeletons changed. With changing diet, their teeth took on a distinctive new form. They were no longer apes; they were proto-men.

An ancient ape, around 20 million years old, experiments with a ground-living existence near the forest's edge. (*British Museum, Natural History*)

The first skeleton in the closeted history of early hominids belongs to a creature called *Ramapithecus*, which lived around 14 million years ago. Pieces of several *Ramapithecus* skulls are known, but anthropologists have not always been in agreement about their interpretation.

The first specimen of this human precursor to be recognized was found in the Siwalik Hills of north-western India in 1934, by Edward Lewis, a young anthropologist from Yale. Lewis believed that *Ramapithecus* was a hominid, but the view of the majority placed it among the apes. This view prevailed for a generation.

In 1961 the indefatigable Louis Leakey, working an ancient site at Fort Ternan in Kenya, unearthed some jaw fragments of an early creature with man-like teeth. This creature had lived 14 million years before, in a forested environment much like that of the Indian *Ramapithecus*. Probably, this whole area of the globe formed a continuous, near-tropical forest at that time.

The Yale anthropologists Elwyn Simons and David Pilbeam saw that Leakey's discovery was another *Ramapithecus*, similar to Edward Lewis's finds. Examining the specimens further, Simons and Pilbeam were sufficiently impressed to slot in *Ramapithecus* as the first firm rung of man's evolutionary ladder.

Though there are no body bones to tell us more about *Ramapithecus*, anthropologists can deduce something about the creature's habitat and diet by the faunal remains associated with the find, and by the nature of its teeth. They build a picture of *Ramapithecus* as a forest-living creature that was becoming adapted to feeding on the ground, chewing vegetable food. This step from the trees to the ground would have been important in the development of early man.

'*Ramapithecus* probably evolved about 15 million years ago, perhaps in Africa,' speculates David Pilbeam. It would then have spread rapidly into Asia, where Edward Lewis's first revealing finds were made.

Although it is classed as a hominid, *Ramapithecus* did not look very much like a man. It was still a small creature, much like a chimp in stature, weighing around 60 pounds. *Ramapithecus* may well have moved among the trees by arm-swinging, an adaptation that was not totally lost even by more advanced hominids. But he probably began to walk upright when he came to the forest-floor.

The change to upright, bipedal walking has been a greater block to our understanding of human evolution than the development of a larger brain. Why should any animal need to walk on two legs, when it could get around well enough on four? Interestingly, many primates occasionally take to two legs. Some, such as the gibbon, may balance along branches on two legs. Others skip across open land on two legs, or stand up to survey the country round them.

The oft-used argument that upright walking wastes energy was swept away in 1972 by two Harvard scientists, Richard Taylor and V.J. Rowntree. In experiments with primates they showed that the amount of energy an animal expends is no different whether it runs on two or four legs.

As man became a hunter, an upright stance would have benefited him. As well as giving him a better view, it would enable him to hold weapons and attack. Carrying things overland needs bipedal locomotion, to free the hands. An explorer who carries his own food can forage farther than the perpetual quadruped who is unable to plan his victualling in advance.

Could *Ramapithecus* have begun to adopt an upright stance?

Possibly it may. To settle this point, anatomists will need more bones to study.

After the tantalizing glimpse of our early history that *Ramapithecus* gives, the mists of time close again across a 10-million-year gap of pre-human history. Surveying the dry, bone-bearing plains of East Africa, a fossil-hunter's fingers may well itch impatiently to dig out the secrets of those intervening jumps up the evolutionary ladder, held in the bands of ancient rock and soil layers. For, during that intervening time, there emerged a new and advanced primate which, in retrospect, we can see had further developed those traits that already set *Ramapithecus* apart from the apes.

This new creature was a pigeon-toed, upright walker with a shambling gait. From five million years or so ago, its type would have been seen living and hunting in small groups on the open African plains, mixing an increasing amount of meat in with its diet of roots and fruit. This relatively hairless, dark-skinned creature, which communicated with various hoots, grunts and rudimentary sounds, would have been recognizable as man in the making. Such a creature would have been no taller than a modern human infant, with long arms that revealed the historical association with his arm-swinging forebears. Yet there now seems little doubt that he was a direct ancestor of ours.

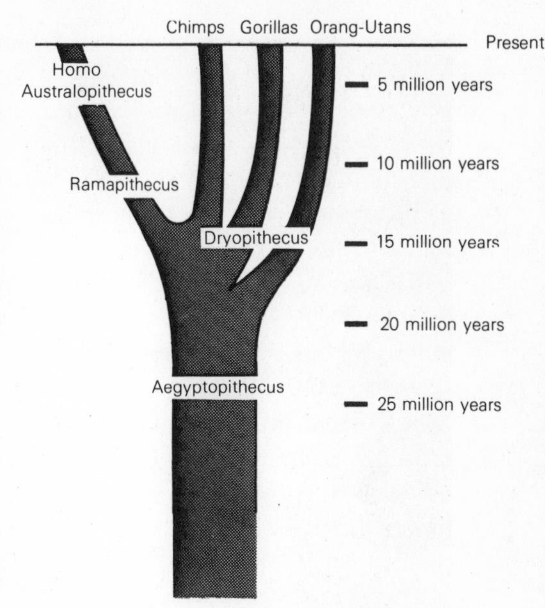

Lineage from early to modern apes. (*Modified from David Pilbeam*)

In 1924 Raymond Dart, then a young anatomist in Johannesburg, cradled the head of an early hominid child, much as its mother must have done after it died at the age of six. He held one of the most important links in our chain of evolution, but at that time few scientists were prepared to regard it as such. To describe this ancient child and its kind, Dart created a new genus which he called *Australopithecus* (literally meaning southern ape), and added the species name *africanus*. The skull had been found by workers in a limestone quarry at Taung, about 85 miles north of Kimberley in South Africa, and was sent to Dart for examination. In the ensuing years, similar fossil remains were to be found at other South African sites whose euphonious names have since become famous: Sterkfontein, Kromdraai, Swartkrans and Makapansgat.

From these specimens, anthropologists have been able to deduce that the adult *Australopithecus africanus* had a brain volume of around 440 cubic centimetres as compared with around 1,400 cubic centimetres for modern man. Clearly, these creatures were intellectually far inferior to modern man, but in them we can see that, in relation to body size, the brain was becoming larger and more complex. 'Contrary to what is widely believed,' asserts Columbia University anthropologist Ralph Holloway, 'the human brain was not among the last human organs to evolve but among the first.' Members of *Australopithecus africanus* were cleverer than modern chimps, for example, although their bodies were evidently no bigger.

The Taung child had not lost his (or her) milk-teeth at death, indicating another similarity with man: a slow rate of growth. And presumably he was surrounded by a family from whom he would learn.

Following Raymond Dart's electrifying claim about the incipient humanity of *Australopithecus*, other specimens were unearthed that served to bolster his view. The sites, which were in ancient caves in sedimentary limestone rock, were difficult to date accurately, but seemed to lie in the region between two and three million years ago. Importantly, clear evidence of the use of tools was found with these remains: actual stone implements, selected and shaped by creatures with the mental capacity and manipulative ability to impose their own

A memorial to Robert Broom, shown holding an *Australopithecus* skull, at Sterkfontein cave, South Africa. Broom was one of the first to support Raymond Dart's claim that *Australopithecus africanus* represented the missing link. In 1935, Broom found remains of *Australopithecus africanus* at Sterkfontein, and later discovered *Australopithecus robustus* specimens at neighbouring sites. (*Ian Ridpath*)

Australopithecus africanus at a South African cave site

idea of form upon the natural world. These stone tools were used to cut meat, pulp vegetables, and possibly even shape animal skins for covers.

Though the deposit sites are termed caves, this does not imply that cave-living men of the Stone Age type had yet evolved. They probably sheltered under rock overhangs, with tropical downpours washing their remains farther back into caves and down crevices, where they became embedded in sediments and have since been laid bare by erosion. The plains of Africa, with their tempting game herds, would have kept *Australopithecus* on the move in the open.

Three fossil sites near Johannesburg — Sterkfontein, Kromdraai and Swartkrans — plus evidence from Makapansgat, showed that *Australopithecus* was widely spread over southern Africa a few million years ago. More importantly, the Kromdraai and Swartkrans sites revealed remains of a second *Australopithecus* type that had emerged. Found originally by palaeontologist Dr Robert Broom in 1938, this fossil primate is now called *Australopithecus robustus*, because of its bulkier nature than the delicate *Australopithecus africanus*.

Australopithecus robustus had larger teeth and stronger jaws than those of *Australopithecus africanus*; evidently it was good at crushing and grinding food in its mouth. Evidence from Swartkrans shows that this creature had a rather larger brain — around 530 cubic centimetres—than *Australopithecus africanus*. But it was probably no cleverer, for its body size was also considerably bigger. These robust australopithecines may have stood as tall as 5 feet, and weighed a considerable 100 to 120 pounds—far larger than their contemporaries.

In July 1959 the scene shifted several thousand miles northwards to Olduvai Gorge, a gaping slit in the surface of Tanzania, where Dr Louis Leakey and his wife Mary were in search of man's fossil ancestry. They found part of a skull that proved that early hominids had not been confined to southern Africa alone. What was more, it was possible to date the specimen confidently: 1.75 million years. The accuracy of this date was a shaft of brilliant light in the uncertain gloom of our prehuman past.

Olduvai Gorge is a side-shoot of the great African Rift Valley that scores its way south from the Red Sea through Ethiopia, Kenya and Tanzania to Mozambique. Here, at one time, were the waters of a shallow lake. This lake had periodically flooded and dried in accord with the changing seasons, coating the animal-remains on its banks with protective silt. Layers of ash, which had been erupted sporadically from nearby volcanoes that flared in sympathy with the shifting ground, laid time-markers between these silt-covered remains like radioactive bands. Scientists can measure the age of volcanic products that have been smelted in the furnace of the Earth's interior, using a technique that depends on the radioactive decay of certain atoms. This was the method employed at Olduvai Gorge, but it is unworkable with sedimentary rocks, like those in southern Africa, that are consolidated from the decay of ancient stone.

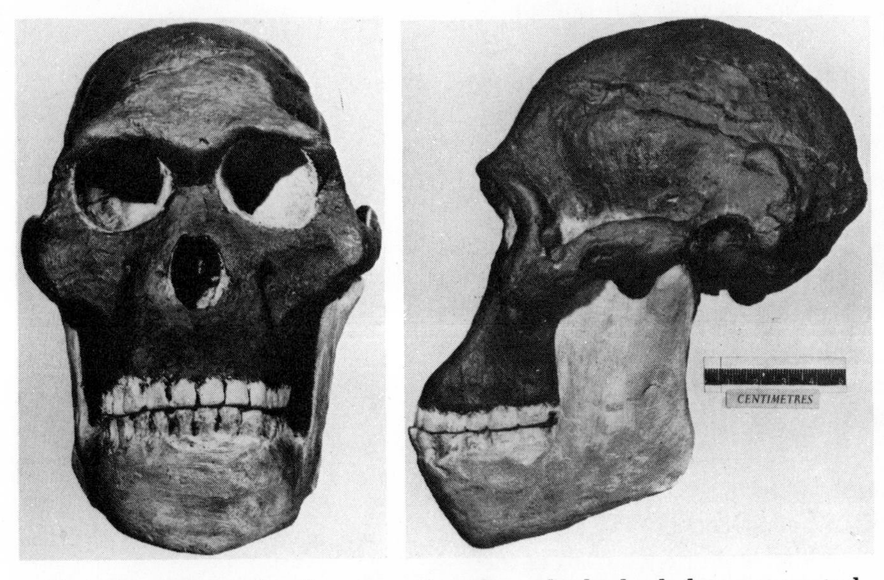

Front and side views of a reconstructed skull of *Australopithecus africanus*. *(British Museum, Natural History)*

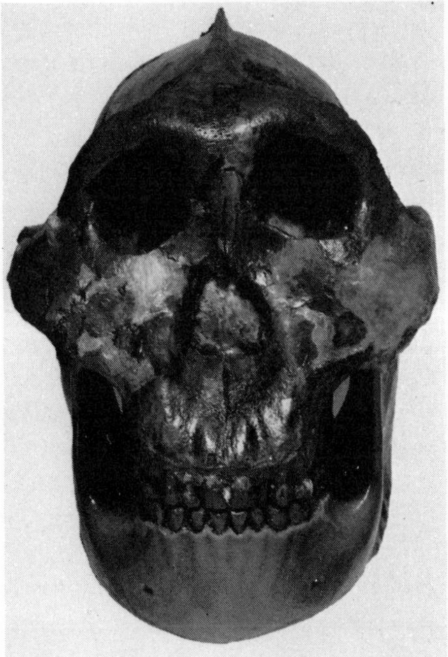

Front view of *Australopithecus robustus*, showing the massive jaw and skull crest to which were attached his chewing muscles. (*British Museum, Natural History*)

Olduvai's hominid remains were accompanied by stone tools made of a rock type not found locally—possible evidence of widespread cultural and physical interchange between the ape-men of East Africa. Scattered round were the smashed bones of cut-up game. A circular pattern of rocks, and the remnants of an australopithecine rubbish pile, told the Leakeys that they probably stood upon the living-site of these early ancestors of man.

To accommodate the new skull from Olduvai, anthropologists constructed a new species, which they call *Australopithecus boisei*. It was apparently an even more strongly built creature than *Australopithecus robustus*—and presumably off the direct evolutionary line to modern man.

Olduvai was to produce evidence even more unsettling to the anthropological establishment two years later. The Leakeys and their co-workers unearthed 1.75-million-year-old specimens of a new type of man, so distinctive that they boldly placed it in the genus *Homo*—implying that it must be a direct forerunner of ourselves. They called it *Homo habilis*, now colloquially known as 'handy-man'.

The evidence for this dramatic step did not convince many other anthropologists. Although the brain of so-called *Homo habilis* was bigger, at around 650 cubic centimetres, it was not yet so large as to place it beyond the possible bounds of *Australopithecus*. Handy-man, who did have a basically man-like type of hand, was still small in stature, probably being no taller than 4½ feet on average.

The question was thrown into wider disarray a few months before Louis Leakey died. His son, Richard, who has assumed his father's mantle as a leading fossil-hunter, discovered at another East African site a skull of apparently even larger brain size than *Homo habilis*, yet living a full 2.9 million years ago. This skull, which bears the catalogue number 1470, was found in fragments during August and September 1972 to the north-east of Lake Rudolf, Kenya. A reconstruction was first exhibited to a somewhat incredulous meeting of England's Zoological Society in November 1972.

Richard Leakey's 1470-man seems to have had a brain capacity of about 800 cubic centimetres, making it far more intelligent than any of its contemporaries, or any of the *Australopithecus* line that lived on for two million years after. 'Brains whose organization was essentially human were already in existence some three million years ago,' says anthropologist Ralph Holloway from a study of the 1470 skull. For Leakey, the skull's importance is that it shows that *Homo* and *Australopithecus* existed together, evolving simultaneously. 'This has never been proved before,' he says. Another Leakey discovery from the evidence at Lake Rudolf is that male and female *Australopithecus* types are built rather differently. This is not a man-like tendency; rather, it is reminiscent of the apes. It serves to bolster the argument that *Australopithecus* was a by-way on the road to modern man.

But the evidence for the time farther back, tracing man's origins towards the era of *Ramapithecus*, remains frustratingly buried in the endless wastes of Africa. Progress is being made in filling the gaps, but it is painfully slow.

From Lothagam Hill in Kenya came the oldest indubitable evidence of an *Australopithecus*-like creature, unearthed in 1967 by a Harvard expedition under Professor Bryan Patterson. This jawbone fragment, probably of *Australopithecus africanus*, has been dated to 5.5 million years.

Other areas, still being searched for fossils, have revealed the existence of tool-making ability over 2½ million years ago in East Africa. But of the proto-men who made the tools there is no sign.

Farther north, in the valley of the River Omo in southern Ethiopia, lies an even more arresting site for fossil-hunters: sediments 2,000 feet thick, covering 2½ million years of pre-history, back to 4 million years ago. Here, numerous *Australopithecus* specimens are being uncovered each year.

'These discoveries have substantially expanded our conception of the time scale for man's evolution,' comments F. Clark Howell, professor of anthropology at the University of California, Berkeley, an avid hunter after fossils in the Omo basin. 'They have shown how complex is the early evolution of the family of man. In the remote past, there were times when several species of men existed at the same time, even in the same area, though only one survived and continued to evolve toward modern man.'

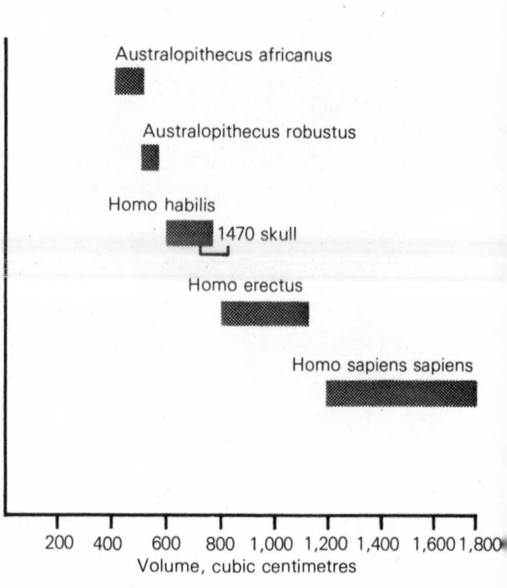

Size range of the brains of various hominids. (*Data from Ralph Holloway*)

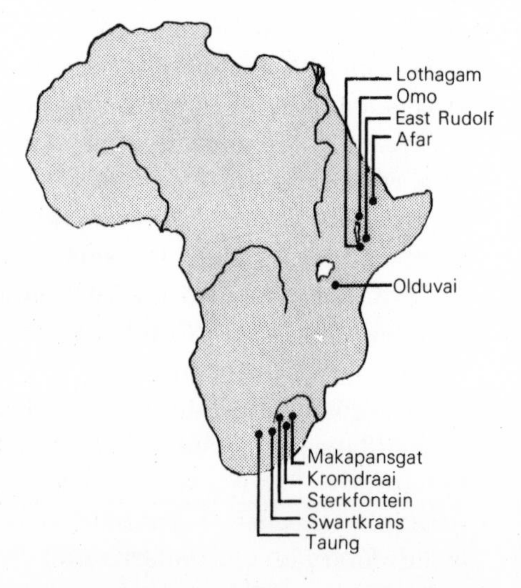

Map of Africa showing sites where major finds have been made of fossil Man.

But which one? And where do the controversial South African hominids fit into this pattern?

In his office at the University of the Witwatersrand Medical School, Johannesburg, Phillip Tobias outlines the various methods now being used to date the South African cave-sites, which allow the fossil men found there to be fitted into a coherent evolutionary picture of early man.

The standard dating technique is to study the animal remains at each site and compare them with similar remains from accurately dated locations. This was first tried in 1970 by Dr Basil Cooke, who compared pig remains from Sterkfontein and Makapansgat with the accurately dated pig remains from East Africa. Since then, other researchers have made comparisons using elephant, rhinoceros and cattle remains.

Says Tobias, 'Gradually a picture is emerging which suggests that the fauna which was contemporaneous with *Australopithecus* in South Africa matches the fauna which was living in East Africa between 2½ to 3 million years ago. This is far older than anyone had dared to postulate for the South African sites.'

Most recently of all, a South African geomorphologist, Dr Tim Partridge, has developed a method based on the erosion of valleys to work out the likely dates at which the fossil-bearing caves were first opened to the surface. The caves were initially dissolved by rainwater out of the underlying limestone, and opened to the surface by the widening of a river valley.

Partridge's dates for the opening of the caves are the most specific of all, and pre-date the ages derived from the oldest known animal remains. This suggests that in most cases there are many hundred thousand years of fossil deposition yet to be excavated, which current investigations by Tobias's group at the Sterkfontein and Makapansgat sites have confirmed.

The major surprise from Partridge's work is that the Taung site is not, as had long been assumed, the oldest of the South African fossil-bearing caves. Instead, it is the youngest—by a considerable margin. The famous *Australopithecus* child from Taung may have lived as recently as three-quarters of a million years ago.

From these new datings, Professor Tobias now draws the family tree for man shown here. He uses the evidence of Richard Leakey's discoveries in East Africa to show that *Homo* first split from *Australopithecus* between 3 and 4 million years ago. And the tree shows that three branches of the *Australopithecus* lineage lived on, side by side, on the plains of Africa, each eventually dying out.

That this area may be open to more drastic revision was shown in October 1974 by a joint French-American expedition to the Afar region of Ethiopia, in the African Rift Valley. There they found parts of a three-foot-high female skeleton, lying under an ash layer dated at around 3 million years old. Though this specimen, named Lucy, remains to be fully examined, the expedition members believe that she shows traits which place her in the genus *Homo*. 'We must consider the possibility that Man's origins go back to well over four million years,' they said. Phillip Tobias declared himself not at all shocked by these revelations. 'It is possible that the specimens may be intermediate between the group I consider to be ancestral, namely *Australopithecus africanus*, and early *Homo* of the 2 to 3 million year period,' he suggested to me.

Fitting the Taung skull into Tobias's family tree of Man remains a

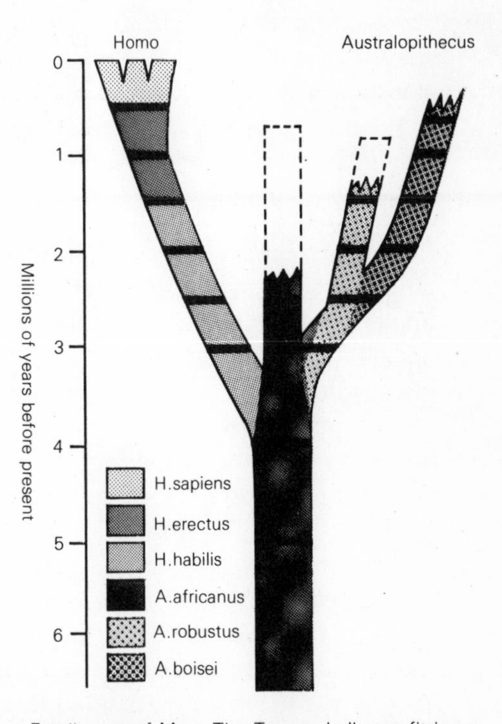

Family tree of Man. The Taung skull may fit in as a late member of *Australopithecus africanus*, or as a member of *Australopithecus robustus*. (*After Phillip Tobias*)

problem. Should it be regarded as evidence that *Australopithecus africanus* survived in isolated pockets over a million years after the main line had died out, while radically different hominids evolved elsewhere? Or should it be regarded as a late example of *Australopithecus robustus*?

Tobias feels that, as the Taung skull is of a child, its true lineage may have been misjudged. He has embarked on a major study of the skull to resolve this doubt.

The new family tree of man shows clearly that the Taung skull and its fellows are examples not of direct human ancestry, but of the evolutionary by-ways that ape-men entered on the rolling plains of Africa. Their brainier counterparts had already side-stepped them on to a new evolutionary track, up whose rungs they nimbly climbed towards the advanced design of modern man.

Tobias sees the differing brain sizes of *Homo habilis* and the 1470 skull as evidence that there was great variety—one might even say experimentation in life-styles—within the populations of the first true men.

The final flowering of that lineage in *Homo sapiens sapiens* was a creature that has proved to be as powerful a modifier of his own environment as are natural forces.

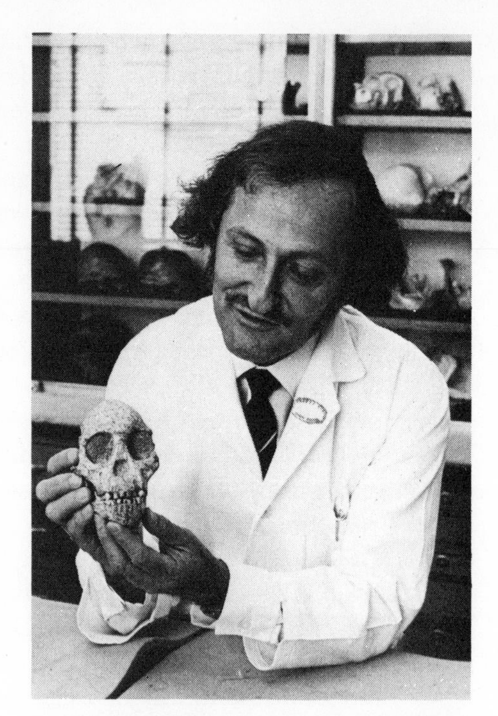

Phillip Tobias and the Taung skull. (*Ian Ridpath*)

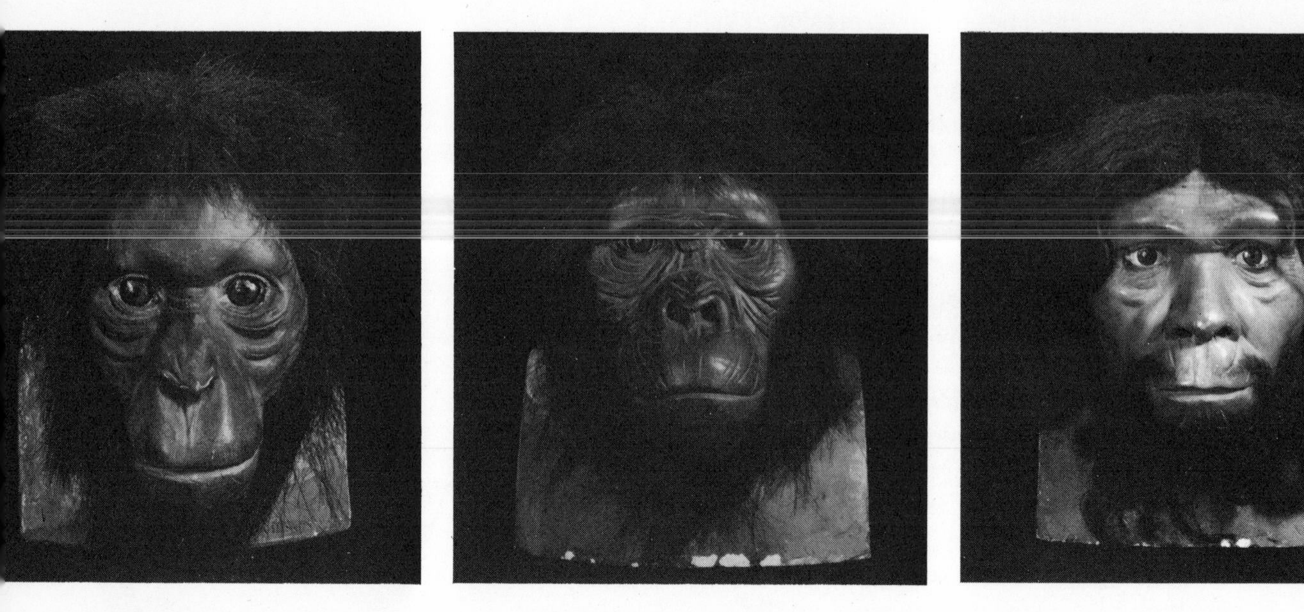

A series of reconstructions, showing the cranial development of hominids from an ape of the type known as *Dryopithecus* (a probable common ancestor of modern apes and men), via *Australopithecus africanus*, *Homo erectus* and Neanderthal man, to *Homo sapiens sapiens* represented by Cro-Magnon man.

When did the first creatures arise that we can term 'people'? The slow flow of evolution crossed this arbitrary line somewhere around $1\frac{1}{2}$ million years ago. That was when men of truly upright stance appeared, walking as efficiently as we do today. We refer to them as upright-walking men—*Homo erectus*. They are represented by the famous Java and Peking types of men.

Java man was the first new species of *Homo* to be found. Its discovery in 1890, in the form of a skull fragment with large ridges over the eyes, came at an opportune time in the history of scientific thinking: though Darwin was dead, his ideas had a powerful influence.

Java man was evidence that man-like creatures, unrefined prototypes of ourselves, were widely spread over the globe a million or so years ago. At the time the find was made, by the Dutch anthropologist Eugene Dubois, scientific opinion was not sufficiently courageous to label Java man under our own genus. So the misleading term *Pithecanthropus*, meaning ape-man, was applied to it. Only much more recently did anthropologists take a deeper breath and venture to put Java man among the ancestors of *Homo sapiens*.

A thigh bone, apparently from the same layer as the Java skull, reveals the remarkable fact of an almost completely modern skeleton in combination with a brain whose volume was only two-thirds that of modern man.

Our knowledge of *Homo erectus* extends far and wide across the globe, and through many hundred millennia of time. Evidently, by about 2 million years ago the precursors of man were spreading from Africa, and we find their descendants in the *Homo erectus* remains of Asia and Europe. *Homo erectus* specimens are also common in East Africa, so this stage in our evolution should be one of the easiest to decipher.

But the wide range of *Homo erectus* brings its own problems, because racial differences start to become evident in each location. Also, evolutionary variations become noticeable over the long time-thread of many of these populations. So the *Homo erectus* tag covers a considerable range of skull forms.

At this time came a major evolutionary force. As men spread from the warmth of their African cradle, the ice cracked down. And, in response, man made the master discovery—fire.

Fire gave man the ability to create his own environment in defiance of the elements, without which he would have had to give ground as a

major ice age swept down on him from the north. But, for the first time, we find that man defied the forces of nature by his own ingenuity. That ability he has been using ever since; and he will continue to shape nature to his own ends until he brings about his destruction, or brings to his aid the guiding influence of more advanced intellects still.

We find that fire was first used on Earth by men who lived in what is now China. Peking man is the common name anthropologists give to extensive fossil-remains found from 1927 onwards at Choukoutien, near Peking.

Evidently, these specimens span the time-range between Java man and the emergence of *Homo sapiens* about half a million years ago. Indeed, the largest brain volume measured for Peking man, at 1,300 cubic centimetres, is hardly smaller than the modern average brain size. Although Java man was apparently around 5½ feet tall, the Chinese specimens suggest a rather smaller stature, at around 5 feet—perhaps an adaptation to the colder conditions.

Here, too, men were living in caves; and the use of fire meant that cooking was becoming the accepted way of preparing meat.

Peking man (*Homo erectus pekinensis*) used stone tools to prepare food, and cooked meat on a fire. (*British Museum Natural History*)

Evidence for even more reassuringly familiar behaviour comes from *Homo erectus* populations sampled in Europe. From a site at Nice, anthropologists tell of *Homo erectus* populations building shelters from stones and wooden poles, roofed with hides, in which they collected round a hearth to cut up and cook food. At about the same time, several hundred thousand years ago, *Homo erectus* hunted big game on the plains of Spain, near what is now Madrid.

Clearly, the line to modern man is now well-drawn. Around four hundred thousand years ago, the changes in *Homo erectus* mount up sufficiently for scientists to refer to the specimens as early *Homo sapiens*. But the story of our rise is not over yet. With world population then stable at round a million, even small migrations could produce major changes in the morphology of men at any one place. The fossil record hints at varying success for different groups of *Homo sapiens*.

The major developments from *erectus* to *sapiens* were an increased brain size, a flattening of the face, and a rounding of the skull outline. During this time, evidently, the brain was reaching its final degree of internal organization, and language was becoming highly developed.

Men were still nomadic hunters, and it was the needs of this way of life that brought about the changes in our ancestors. Creatures that could plan ahead, co-operate and communicate would have the greatest hunting success. Some contend that the highest intellectual abilities—of aesthetic sense and cultural creativeness—that we value so highly are accidental by-products of the need to survive.

For all of its relative recency, the evolutionary path from the first

Artist's impression of a Neanderthal family group at Gibraltar about 75,000 years ago. (*British Museum, Natural History*)

Homo sapiens to contemporary human populations has not been easy to trace. For instance, fossil remains from a later population of men than *Homo erectus* have been found from the Solo River in Java. These are probably the descendants of Java man, for they show certain *erectus* characteristics. This evolution evidently proceeded without much crossing of genes from elsewhere, and there are signs (though slight) that the Solo line progressed to present-day Australian aboriginals.

Elsewhere, there are examples of local populations being totally deposed by the influx of more-advanced migrants. This was the fate of Neanderthal man, a rather specialized form of *Homo sapiens* which was entering an evolutionary dead-end about sixty thousand years ago. The neanderthal's brain was, if anything, larger than that of modern man, but the skull was flatter and the face, along with many parts of the body, was generally bulkier than average today. He was of the type we might term a 'bruiser'. The neanderthals flourished in western Europe, but do not seem to have been important in the development of contemporary man. Sir Wilfrid Le Gros Clark described the Neanderthal type of man as 'an aberrant side-line of evolution, the result of a sort of evolutionary regression'.

The evidence of genetics combines with the archaeological story to confirm that, around 40,000 years ago, advanced types of *Homo sapiens* spread from the east into Europe to produce today's populations of *Homo sapiens sapiens*. This pattern was repeated in many parts of the fossil history of *Homo sapiens*. In other words, it seems that although *Homo erectus* spread throughout the world, and developed at various points into different *Homo sapiens* species (itself an interesting point), there was later another wave of more-advanced *Homo sapiens*. These new migrants seem to have provided the major genetic basis for modern man, but the picture is confused by the inevitable crossings that occurred with pre-existing *Homo sapiens* populations.

Although they believe that they can recognize the effect of this new wave of men throughout the world, anthropologists still puzzle about its geographical origins. The available evidence points to somewhere in North Africa or the Middle East as the most likely spot. On Mount Carmel in Israel, caves have yielded examples of skulls, perhaps fifty thousand or more years old, that clearly anticipate modern man in many of their features. Even earlier examples of modern-looking men are known from layers dating back perhaps as much as a hundred thousand years in the Omo river valley of Ethiopia.

Since the emergence of these modern-looking forms of men there have been no major advances in brain structure; rather, the superior types have prospered at the expense of the inferior.

Says David Pilbeam, 'The time between 50,000 and 30,000 years ago saw the spread of modern man out of his hypothetical "garden of Eden" until, through a process of swamping and replacing older and more archaic subspecies of *Homo sapiens*, he inherited the Earth.'

Can we expect that any form of life even remotely resembling man would arise on other planets? Biologist Harold Blum has attempted to estimate the number of genetic mutations and cultural innovations involved in the evolution of man. 'We reach a total figure of 10^{-18} [one in a million million million] for the probability of man having reached his present state,' Blum concluded, adding that this is the rough inverse of the number of likely life-bearing planets in the visible Universe.

World-renowned zoologist George Gaylord Simpson has also spoken of what he sees as the 'almost negligible' chance that life on other planets would evolve to organisms with human-like intelligence and capabilities. He calls the assumption that life will lead to humanoids of some kind 'plainly false'. For Simpson, the continuing diversity of the rate and direction of evolution, with many types of creatures becoming extinct, shows that there is no goal-directed route from single cells to intelligent man. With only slightly different starting conditions, he maintains, the end result would be profoundly different. The enormous diversity of life today represents only a small fraction of what is possible. 'Evolution is not repeatable,' says Simpson, pointing out that no group has ever evolved twice. 'Any close approximation of *Homo sapiens* elsewhere in the Universe is effectively ruled out,' he claims. 'I therefore think it extremely unlikely that anything enough like us for real communication of thought exists anywhere in our accessible Universe.'

And in a special message to the participants at the Byurakan conference on extraterrestrial intelligence, he underlined the fact that the continual pushing back in time of life's origin 'decreases the probability of a parallel origin of intelligence elsewhere.' In a clear censure of the conference's distinguished participants, he said, 'Exobiology (the study of extraterrestrial life) is still a "science" without any data, therefore no science.'

Not all biologists agree with Simpson's argument. Instead, they draw attention to the phenomenon termed parallel evolution, in which animals of widely differing stock begin to converge on the same design.

An example is the invention of flight by a wide range of creatures from insects to birds, mammals, reptiles and even fish.

'There is no question that there are a great many individually unlikely steps which led to the development of our technical civilization,' says Carl Sagan. 'But are there not many other sequences of steps that would lead to a more or less equivalent civilization?'

Harvard anthropologist William Howells has pondered the likely nature of people on other planets. His thesis is that, once one assumes that they are intelligent, they must also be 'human', in the sense that they have culture, like ourselves, and that they communicate ideas to one another, and create things jointly: 'Otherwise,' he says, 'intelligence means nothing.'

People on other planets would, thinks Howells, have structures analogous to our limbs and sense-organs. To be of most use, the sense-organs will probably cluster near the main nerve-centre, the brain, at the being's front end: 'Therefore our men will have heads.'

As we had to be able to *do* things to become human, both in the sense of movement and manipulation, Howells directs: 'Look for plenty of fingers on the ends of two arms.' As for legs, he points out that the fishes once had a larger number to choose from. 'I will lay a small bet that the first men from outer space will be neither bipeds nor quadrupeds, but bimanous quadrupedal hexapods' (a term Howells invents to convey a six-limbed beast that walks on four legs). 'If they have four feet to hold them up, then they might well be as big as a horse, or larger,' he says.

Robert Bieri, a biologist at Antioch College, Ohio, believes that the environmental conditions needed for the formation of life impose 'severe limitations' on the number of evolutionary routes available to living things. 'If we ever succeed in communicating with conceptualizing beings in outer space, they won't be spheres, pyramids, cubes, or pancakes,' Bieri declares. 'In all probability they will look an awful lot like us.'

Howells asks himself whether men would rise again if our species were suddenly extinguished. Primate specialist John Napier has said that macaques and baboons might well have been the dominant forms of animal life in all the temperate regions of the Old World—were it not for the coming of man. But Howells is not impressed by their chances of taking over from man because they have made no progress in the past 35 million years. And the apes, he suspects, are already too specialized to change. 'The next try would have to come from a tree shrew, laboriously repeating all of primate history,' he says. Yet there would be hard competition from the existing higher mammals, so perhaps these should also be swept away. 'All in all, our hopes of repetition are not good,' Howells concludes.

Primates apart, what other animals could respond to the challenge if the human role, or niche, in the natural world were vacated? Musing on this point at the Byurakan conference, Philip Morrison announced that one type of animal that looked favourable was 'the omnivorous, hand-using, cute, and responsive raccoon'.

Marvin Minsky, a researcher into the origins of intelligence and a colleague of Morrison's at the Massachusetts Institute of Technology, added the whale, dolphin, and cephalopods (a group including the octopus and squid) as additional candidates.

For those who doubt that mankind is unique, good news comes from anthropologists, such as Phillip Tobias, who are realizing that the emergence of intelligent, creative hominids on Earth was not as unprecedented as has hitherto been supposed. Evidence is mounting that around the time of *Ramapithecus*, other creatures were developing some of the ways of man, although not all together. Only in one type did all the best features—such as locomotion, manipulation, and co-ordination—come together to ensure its survival and development towards man.

Tobias sees a more recent parallel of this evolutionary pattern in the emergence of the larger-brained hominids in East Africa whose skulls have been found by Richard Leakey. Says Tobias, 'I believe that intelligence is the *sine qua non* of our type of mammal. Everything has come to depend upon the brain and what we do with it.'

In the days of the early hominids, when youthful death was the norm—half the African specimens of early man died before they were fully mature—living long enough to have children was a full-time job. So a creature with enough extra wit to live a few years longer would have a major influence on future generations because of the extra children he would leave behind. At that time, the selective advantage of a bigger brain was evidently enormous: a fact which the apparent rapid development of brain size underlines.

Says Tobias: 'The view that mankind's development was a lucky chance, and the only one, may perhaps be not quite right. It may well be that nature was making a number of experiments in hominization, and the one line that developed from early *Homo* to *Homo sapiens* had the total concatenation of these complexes which made for such advantage that it survived and ousted all other hominids.

'Man is unique in his total combination of all the features, but any one of them has been tried before.

'It's quite conceivable that, given the same starting conditions, and given enough time and evolutionary opportunity, it could happen more than once.'

3

THE FUTURE OF MAN

The evolution of mankind is far from over. Even today, subtle changes are still traceable in the molecules of life that make up our bodies.

Whether the final result will be better or worse we cannot yet tell. Could we be reducing our biological capacity for survival?

In our modern civilization, man-made environments have ended the need for adaptation, and care for the sick has removed nature's way of dispensing with unfit genes. Selection is no longer natural; we have overcome evolutionary forces by our humanitarianism. And it is difficult to imagine any advanced society doing otherwise. So Sebastian von Hoerner's view, reported in Chapter One, that some civilizations decay through genetic deterioration, may one day become a depressing reality for ourselves.

However, there are two evolutionary forces that may yet change our race beyond recognition. One is self-inflicted biological engineering; in other words, the ability to change our own genetic material in the laboratory, and create or destroy hereditary traits at will.

The late J.B.S. Haldane had a favourite idea that the prehensile tails of monkeys made them better adapted than man for moving around in weightlessness—and that gene grafting might make it possible to incorporate such a useful extra appendage in human stock. This seems a trivial change compared with the power of the method that will make it possible.

The prospect—both horrifying and fascinating—of bio-engineering man is brought ever closer by scientists who probe into the working of cells. The day when we can produce test-tube babies to order, like custom-built cars, remains just a hazy vision at present. Yet we must realize that it can become possible.

The consciences of scientists, and the awareness of legislators, have not long awakened to the consequences of our actions that have hardly been examined outside the pages of science fiction. Says Alvin Toffler, 'We can now imagine re-making the human race not as a farmer slowly and laboriously breeds up his herd, but as an artist might, employing a brilliant range of unfamiliar colours, shapes, and forms.' Others go even further—to producing mutations of man and machine. 'Space biology is marching irresistibly towards the day when the astronaut will not merely be buckled into his capsule, but become part of it in the full symbiotic sense of the phrase,' says Toffler. Space engineer Theodore Gordon suggests, 'Perhaps it would be simpler to provide life support in the form of machines that plug into the astronaut.'

Faced with extraterrestrial life of this sort, we might have to decide whether to speak to it or tap a keyboard.

A plug-in astronaut is akin to the cyborg concept of the Americans Manfred Clynes and Nathan Kline, which space chronicler Mitchell Sharpe calls 'the ultimate in man-machine combination'.

In his book *Living in Space** he writes:

> The cyborg is ideally suited for space exploration because it can adapt to a variety of alien environments, but yet retain its Earth organism heredity. In other words, man is modified mechanically and chemically to function in a deadly physical environment such as the atmosphere of Venus or Jupiter. When his mission is complete he can be modified again to regain his original form and function.

The harbingers of the cyborg human are with us in society today—the citizens with electronic pacemakers for their hearts, artificial valves and piping in their bodies, and metal and plastic joints and limbs.

Even more radical is the suggested use of linked human brains as the central processing elements of computers, electrically connected to mechanical peripherals. There seems no medical block to keeping brains alive in isolation from a body. Said one doctor engaged in this research, 'We could keep Einstein's brain alive and make it function normally.' Here also is a prospect of immortality first glimpsed by J.D. Bernal: a bank of brains, each renewed as it wore out, retaining human consciousness for ever, like one eternal individual.

Beyond this, there is the prospect that all aspects of humanity will be supplanted by machines. In their book *Intelligence in the Universe*,† Roger A. MacGowan and Frederick I. Ordway speculate that 'within a very few years it will be possible to build an artificial automaton having superhuman thinking ability, as it is now possible to build a machine having superhuman mechanical strength.' The authors add: 'It is conceivable that intelligent automata may be much more widespread in the Universe than intelligent biological societies.'

Such super-brains are going to have an unimaginable impact on the development of our civilization. But that is beyond the realms of purely biological evolution. Let us first look at one other biological evolutionary force—space itself.

Arthur C. Clarke has compared the evolutionary importance of man's first forays into space with the moment when fishes crawled out of the sea. This colourful analogy is in some ways an inversion of the truth. For, moving into weightless space where he has to take his own air, man is almost moving back into a fish-like environment. Floating fishes (or aquanauts) are as weightless as orbiting spacemen.

Space wreaks swift changes in those who fly weightless. Blood, bones and bodies of astronauts have shown remarkable adaptations to weightlessness after only a few weeks in space. On longer missions, such as the several months spent in the Skylab space station, astronauts have had to exercise daily to prevent wastage of their muscles. Legs weaken without the pulling force of gravity, and hearts pump less hard to push blood round the body. Several weeks have been needed for the astronauts' bodies to re-adapt themselves to Earth's gravity—some have even been unable to walk properly at first.

Half-stone weight losses and one-inch reductions in calf-muscle measures were reported for Skylab crews, despite their exercise programmes. Each member of the third crew, which spent eighty-four days in orbit, came back about two inches taller. As a space doctor explained, 'Up there you find your joints ease up and your muscles flex and your back straightens up. You don't carry yourself around like you do on Earth and you tend to lengthen out a bit.'

Apart from the long-faced look, there are other evolutionary

* Aldus, London, 1969.

† Prentice-Hall, New York, 1966

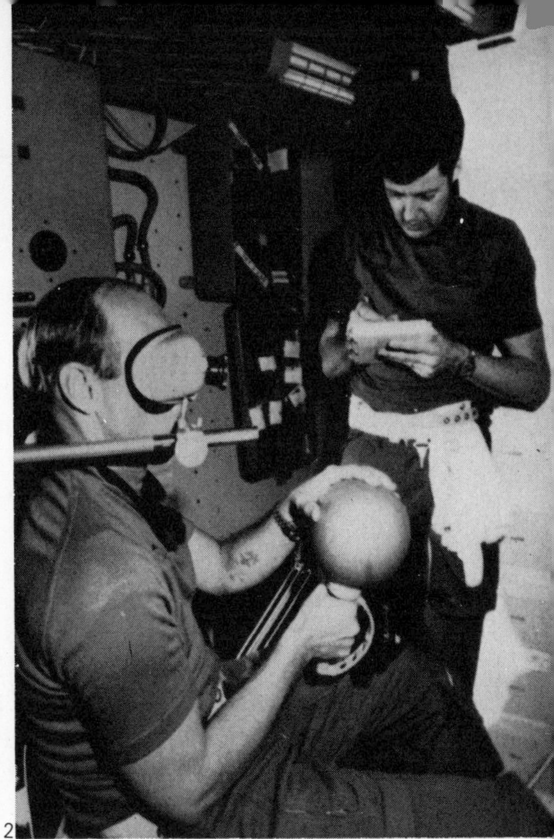

1 Medicine in space. Skylab medical doctor Joe Kerwin puts fellow astronaut Paul Weitz through an experiment to test the ability of the heart to pump blood after prolonged weightlessness. The lower half of the test subject's body is subjected to reduced pressure by suction equipment, placing a strain on the body's circulation. (NASA)

2 Psychology in space. Skylab doctor Joe Kerwin notes the performance of crew commander Pete Conrad in a test to measure visual disorientation in space. (NASA)

3 Exercise in space. A television picture sent back from Skylab shows astronauts Pete Conrad, Paul Weitz and Joe Kerwin preparing for a sprint round the inside of their cylindrical orbiting workshop. Starting from a crouching position, they gained enough speed to run upright, keeping their feet firmly pressed on the inner walls of the space station and creating an artificial sense of gravity. (NASA)

changes of far greater consequence that space-living might bring about in us, and which we may encounter in other space beings, if there are any. In April 1974, not long after the third Skylab crew's return, Soviet geneticists reported on several years' study of animal and plant cells which had been in space. Their conclusion—that weightlessness may change heredity—was reported by Nikolai Dubinin, director of the Soviet Institute of Genetics. Mustard seeds that had been into space produced plants with colours that ranged from green to brown. The Soviet scientists attributed this change to the influence of weightlessness.*

A writer in the magazine *New Scientist* once suggested, not altogether in jest, how the physiological changes brought about by space could be used to advantage: 'The heavy losses of bone calcium

* However, the U.S.S.R. Academy of Sciences has since recommended that Dubinin be removed from the directorship of the Institute of Genetics.

from astronauts suggest that the skeleton and limbs of space-adapted creatures would soon atrophy, and a race of fish-like space cows and sheep would evolve, with a much higher proportion of meat and much more uniform areas of leather and wool.'

More seriously, there are medical hazards to long-term spaceflight that even the science-fiction writers seem not to have considered. The earliest ideas—that men would be sterilized by radiation and peppered by micrometeorites as soon as they peeped above the Earth's atmosphere—are scares that have thankfully been disproved. But there are radiation hazards, and they have an insidious effect.

Cosmic rays are energetic particles that scoot round space, in many cases apparently ejected from the supernova explosions of giant stars. We are blanketed from these by the atmosphere, but in space they riddle the bodies of astronauts. The Apollo Moon crews reported seeing flashes in their eyes as these particles drilled their skulls, leaving scarred tracks through the plastic of their helmets, which have allowed scientists to make a count of the cosmic-ray influx. The results are alarming. As the cosmic rays streak through the head, long chains of destroyed brain cells are left. One estimate is that 0.12 per cent of an astronaut's brain could be destroyed during a two-year trip to Mars, as well as parts of his eyes and nervous system.

This roughly doubles the natural decay rate of human brain cells, so that an astronaut on a long-distance trip would grow senile at twice the rate of his stay-at-home brethren.

No astronauts have suffered sterility problems back on Earth after space missions, and many have sired normal offspring.

To date, only one woman has flown in space. She was the Russian, Valentina Tereshkova, who remained in orbit in Vostok 6 for three days in 1963. Later, she married the Soviet astronaut Adrian Nikolayev, and gave birth to a healthy daughter.

It seems that it will not be long before more women fly in space. Former NASA administrator Dr Thomas Paine once said, 'In the future we will want to have both men and women from many nations working in our space stations.' Said NASA space-medicine expert Dr Charles Berry, 'The lack of normal sex relationships could cause a significant build-up of tension during a mission lasting a year or more. For missions such as the Mars type, mixed crews must be seriously considered.'

In 1973, the United States made studies on a group of would-be female astronauts, and had to conclude that they seemed medically fit for space. In the past, NASA's refusal to counsel the idea of women astronauts brought strong women's-lib reaction. They pointed to the Soviet Union's pioneering woman in space.

With his eyes on future space co-operation with the Russians, Dr Berry mused at a space-medicine conference in Nice, 'The idea of an American man and a Russian woman marking up a joint first in space would really set the seal on our cooperation.' But, shortly after, word got out that the U.S.S.R. had quietly dropped its females-in-space programme.

The peculiar conditions of weightlessness promise to change the whole trigonometry of sex. For one thing, any astronaut knows that pushing against something in space sends you sailing backwards. What will be the means of overcoming (and making the most of) these peculiarities? I look forward with interest to the first instruction manual for sex in a weightless environment.

There are more serious drawbacks. Doctors must be apprehensive about the prospects for the first space pregnancy. All mothers-to-be

Working in space. Apollo 17 astronaut Ronald Evans retrieves a film canister from the side of the spacecraft while returning from the Moon. (NASA)

are advised to keep well clear of radiation, such as X-ray machines, for fear of damaging the foetus. Space is riddled with X radiation, plus radiations of all other wavelengths—including cosmic rays. A developing embryo might be effectively mutilated by a direct hit from a cosmic ray, or steady saturation by X rays, gamma rays, and all the other forms of radiation. It is possible that a space pregnancy could result in horrifying deformity of the foetus.

But supposing that we shield the living-quarters of family spaceships (which might in any case be necessary to reduce the cosmic-ray and solar-radiation flux on adults), we are still left with the major unknown effects of weightlessness on children born in space. No one can answer when we ask, 'What will be the fate of any embryo that develops totally weightless?' We might guess that its bodily strength will make it too weak to withstand high gravitational forces. It could be a kind of jelly baby.

The first child to be born in space may find itself for ever cut off from its parents' planet. Even if it could withstand the full force of gravity, the stress of re-entry might prove too much for it.

Artificial gravity could provide an answer to the problem of weightlessness in spaceships. Indeed, it may be necessary to provide artificial gravity on long-stay space stations, to prevent too much physical decay among the inmates.

But, on the Moon and other small bodies, increasing gravity artificially is not possible. The colonists will have to put up with the weak field.

Moon gravity is one-sixth that of the Earth; Mars gravity rather more than one-third. Someone coming back from Mars would have to get used to weighing nearly three times as much as on that planet. But a child from the Moon would find his weight increased six times.

Even trained astronauts encounter 6g only briefly, during launches and re-entry. What would be the effect on a human who experienced 6 g for long periods of time?

High-gravity experiments in centrifuges show that animals do not live long when taken to a high-gravity regime. Such experiments are equivalent to the arrival on Earth of a creature reared in weightlessness. Gravity changes can be adjusted to in small stages. Mice under slightly increased gravity first showed some sluggishness in feeding and movement, but later behaved normally. Breathing efficiency was also reduced under high g.

These are physiological changes, but over a time they will isolate certain individuals to create a gene pool, which can then lead to genetic change—as when apes and men emerged from the same stock.

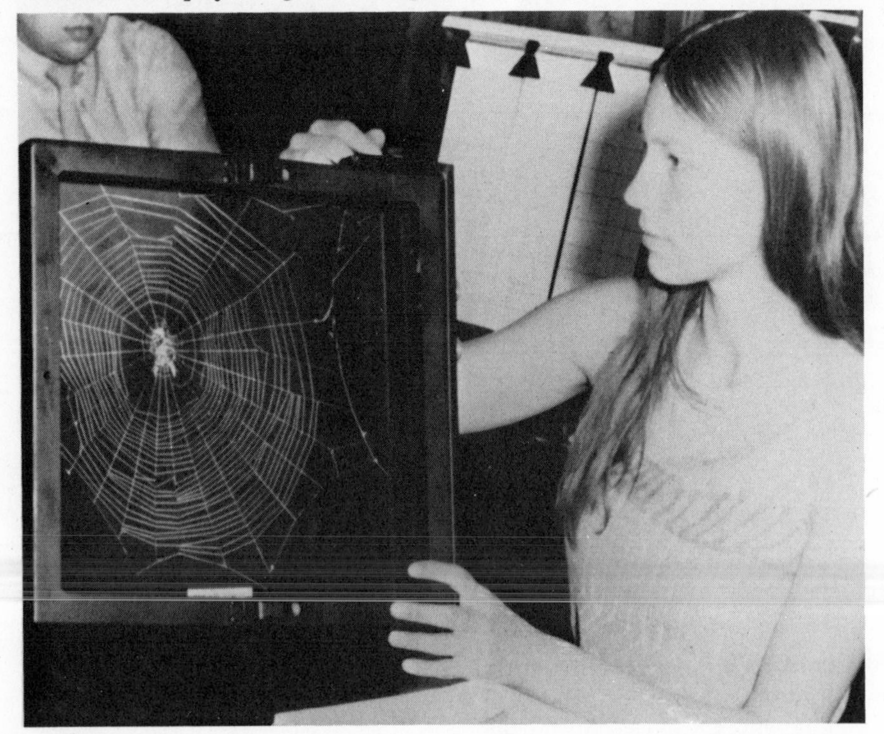

High-school student Judith Miles suggested this experiment to test the ability of a spider to spin a web in weightlessness. After one or two attempts aboard the Skylab space station, two female spiders, Arabella and Anita, successfully adapted to the conditions and began to spin perfect webs inside their test frames. (NASA)

In their book *Planets for Man*, * Stephen Dole and Isaac Asimov termed space flight 'a new form of evolutionary pressure' from the fact that those selected for each new stage in space exploration will have exceptional qualities.

They also imagined a colony isolated on a 1.5 g planet. There would, they concluded, be a premium on bodily strength and higher speeds of reaction in such a weighted-down environment. The best survivors under such conditions would be the most compactly built humans: short in stature, thick of bone and strong in muscle.

Yet the opposite would be true on a planet with gravity lower than 1 g. If the atmospheric pressure were lowered as well as gravitational force, individuals with larger respiratory intake would be favoured.

Such populations, once acclimatized to their new planet, would accept more extreme conditions on other planets. They could then widen a little further the boundaries of what we class as 'habitable'.

Asimov took this idea further in his 1974 book *Our World in Space*. †
In this he imagined three varieties of human beings living in high g, low g and zero g environments, none of them keen to live with the others for long periods of time.

It seems, therefore, that our remote descendants will have very different ideas about what they term a 'comfortable' environment.

* Methuen, London, 1965.
† Patrick Stephens, London, 1974.

Artist's impression of a suggested giant space station in which up to twelve men will live and work for long periods in space. Such space stations may be established in orbit round the Earth during the 1980s. (NASA)

We shall move out into space just as we moved across the Earth. We shall not only colonize the Solar System — we shall conquer it.

What we do in space will depend both on what we find there and on our attitudes. On our own planet, we boldly mount engineering schemes that draw off the natural benefits of our world, and subtly reshape the landscape round us. This will not stop once we reach beyond the atmosphere. In fact, with a larger canvas to work upon, our schemes will become bolder. We shall engineer on a massive scale in space: astro-engineering.

Such exploits may leave a mark on the Universe that will help signal our existence to other beings. Even without embarking on such schemes ourselves, we can look for signs of them being done elsewhere.

One of the first scientists to suggest a major astro-engineering project was Freeman Dyson of Princeton's Institute for Advanced Study — where Einstein spent his latter years of research. In his now-famous proposal, Dyson foresaw that an advanced civilization, pressed for a source of energy to power the needs of its burgeoning masses, might build up a giant shell round its own star to trap the outgoing light and heat. This star-shell would act as a collector of solar energy and, in Dyson's words, 'could contain all the machinery required for exploiting the solar radiation falling onto it from the inside'.

This would be a logical step for an advanced technological group to take. At present, far more energy falls as sunlight on to our planet than we need. Yet, in a few thousand years, our requirements could have

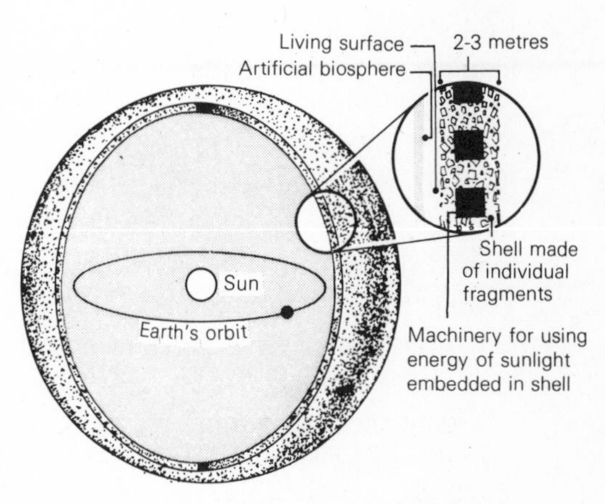

Artist's impression of a Dyson sphere, made of fragments each individually orbiting the Sun. Inside the shell would be machinery for making use of the trapped energy from the Sun. On the inner surface of such a sphere could be a habitable region, or biosphere. Seen from outside, such a sphere would look like a large, cool star.

grown a million million times. Harnessing the total power output of the Sun would be the most efficient way of supplying such energy levels.

The concept is now referred to as a Dyson sphere. To build such a sphere, Dyson anticipates that we should need to disassemble a planet, perhaps Jupiter, and put its fragments into orbit round the Sun. They would continue to move for ever, like tiny planets. The shell of fragments would be several feet thick, and placed about twice as far from the Sun as the Earth's orbit.

Inside this cosy shell, totally new living conditions could be created. The inner region of the shell could be turned into a habitable biosphere for the massive population of the future world; for certain, a planet with a soaring population such as ours would soon run out of land on which to live.

Such a giant sphere would heat up as it absorbed the energy from its parent star, and would re-radiate this heat to space. How would such an object appear from another solar system?

We would not be able to see the star, because it would be hidden by its dark shell. But that shell, sending out heat radiation, would be detectable by astronomers with infra-red sensors. They would measure the re-radiated heat. And they would be able to work out the size of the emitting source. To an astronomer, therefore, such a civilization would appear like a large, cool star.

Is there any observational evidence for such objects? The Cornell University astronomer Martin Harwit has discussed existing infra-red surveys of the sky with this in mind. If Dyson's ideas are right, we would expect to find infra-red emitters with temperatures roughly that of air temperature on Earth. The maximum emission is at a wavelength of 10 microns. Though several likely candidates have been found, the astronomers who performed the surveys believe these to be large stars masked by naturally occurring clouds of dust and gas.

There is no known object that can be explained only as a Dyson-type civilization, and Harwit concludes that: 'other criteria will have to be formulated if a search is to distinguish these civilizations from naturally occurring infra-red astronomical objects.'

Another major form of astro-engineering activity, equally bold in concept if not on such a scale, involves turning the planets into habitable zones for Earth life. Carl Sagan terms this *terraforming*. He first suggested it as a possible way of taming the heat of Venus, and moulding its poisonous atmosphere into something humans could breathe.

Venus, the second planet from the Sun, is surrounded by carbon dioxide gas, which traps heat from the Sun, forcing up temperatures to an oven-like heat on the surface. Astronomers once speculated on what lay beneath the unyielding white clouds that top our view of Venus: tropical forests or sparkling sea water were two ideas. But since then, spacecraft have swooped in close and sent us their news: 'Too hot to handle.'

Venus rocks glow cherry red with the heat. We cannot breathe the scorching gas of its atmosphere. But, though it holds little promise for colonists, not all the options are closed. 'There is a bare possibility of re-engineering Venus into a quite Earth-like place,' says Sagan.

By his own admission, this is a formidable task indeed. Yet the possibility exists because, in with the carbon dioxide, is a damp breath of water vapour. If this were squeezed out of the atmosphere it would make a layer of liquid water over Venus only one foot deep; but that would be sufficient for this scheme.

In an idea that he first proposed in 1961, Sagan foresees that we could prepare simple organisms, such as blue-green algae, for survival in the upper layers of the clouds of Venus. An interplanetary fleet would seed them into the clouds, where they would begin to transform Venus into a likeness of the Earth.

How do we get algae to work for us in this way? These micro-organisms break down carbon dioxide and water, forming free oxygen. The process is photosynthesis, and it gave us the free oxygen on Earth that allowed more advanced forms of life to develop six hundred million years ago.

The windy gyres in the Venus atmosphere will swallow some algae below the clouds, which will fry in the heat, but their decomposition itself will release new water to replace that which has been lost. We must arrange that the algae photosynthesize at a quicker rate than they are roasted. Slowly, the atmosphere on Venus will clear. The insulating layer of carbon dioxide will break up, and oxygen will become abundant. We shall have breathed new life into Venus.

Yet what we see will not be a very rich planet. Its surface will lack running water, and will bear no soil. Where streaking winds have filed the rocks into tumbling fines, the steady heat will have welded them solid again.

Even a terraformed Venus will be no haven for immigrant farmers.

If we go to the planets, it will not be a solution to our population problem on Earth. That must be solved at home.

Terraforming schemes presuppose an understanding of the planets we are about to modify. Soon, cosmic conservationists will be crying for the retention of our planets as they are.

We face moral problems if life already exists on the planets we wish to terraform. Mars may be found to harbour its own organisms that would die in changed conditions. Here might lie the clue to life's origin — and the seed of great civilizations to come.

If life is found, Sagan would be the first to say, 'We should never perform such terraforming.'

The terraformers have their eyes on the gases locked up as frozen deposits in the polar-caps of Mars. Carl Sagan first drew attention to these in 1971, in a paper which put forward a dramatic scenario of changing climate on Mars. Working out what would happen if all the material from the north polar-cap were vaporized, Sagan came to an arresting conclusion: 'There would be as much atmosphere on Mars as on the Earth.' Yet under present-day conditions the north polar-cap does not evaporate totally, and the reason for this is the mis-shapen orbit of Mars.

When the north-polar region of Mars is tilted towards the Sun, Mars is too far away in its orbit for this area to get very warm; and when Mars is closest to the Sun, the north pole is pointing away. So the northern region of Mars remains in a permanent winter. But this will not always be so, and this is the key to Sagan's idea.

Mars wobbles slightly in space, like a slowly gyrating spinning-top. The axis of Mars, therefore, does not point constantly to the same direction in space. One complete wobble takes around fifty thousand years, during which time the north pole comes to face the Sun when Mars is at its nearest.

By then, as Sagan pointed out, the gas will have been transported through the atmosphere to the southern pole and the roles will be reversed. But, in between, neither pole points away from the Sun at the nearest part of the orbit, and the material of both polar-caps

The north pole of Mars, photographed by the American Mariner 9 space probe. Even in the northern hemisphere summer on Mars, this cap does not completely evaporate. If we could make it do so, the climate of Mars might be transformed. (*Jet Propulsion Laboratory*)

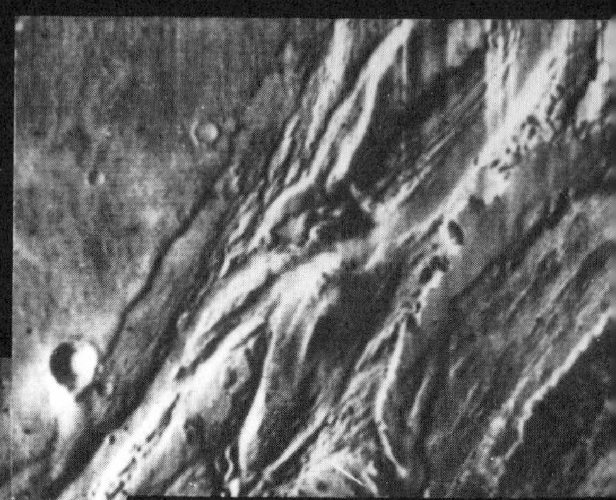

This Mariner 9 picture shows that free water ma once have flowed on Mars — as it could again i we melted the polar caps. (NASA)

evaporates at the same time. When that happens, the climate of Mars is transformed.

The carbon dioxide gas locked away in the polar-caps coats the planet in a warming blanket, and the previously frozen water flows as rivers over the surface. Simple organisms could flourish, before going into hibernation for the coming long winter.

Space-probe photographs of apparent dried-up channels on the surface of Mars lend considerable credence to Sagan's speculative Long Martian Winter model.

Two Cornell University scientists, Joseph Burns and Martin Harwit, dwelt in 1973 on the possibilities of engineering the motions of Mars so that the thick gases were always free in its atmosphere. Mankind might then be able to watch the growth of underdeveloped biological organisms that may be lying latent in the Martian soil. Burns and Harwit looked for ways to unbalance the wobble of Mars so that the warm, damp springtime conditions were maintained. They calculated that moving the Martian moon Phobos closer to the planet would produce the desired effect. Equally, some asteroids could be brought close to Mars, possibly strung out in an attractive ring round the planet.

Such schemes, Burns and Harwit said, are likely to have been used by other civilizations, for the long-period wobble of Mars is not unique among planets; the Earth has such a motion, though it does not affect our climate. 'There is always something a little repugnant about man pushing his own interests and fixing nature,' they wrote in the scientific magazine *Icarus*.

But man has time enough to come to terms with his meddlesome nature: the next Martian spring won't be along for 10,000 years yet. By then, Burns and Harwit believe, the relevant technology will be available—if man is still alive.

A cheaper and easier solution was proposed in 1973 by Sagan himself. If the polar-caps could be made to absorb more heat from the Sun, they would melt. One way of doing this would be to scatter them with dark dust. To seed Mars with sufficient dust from Earth, or to crumble an asteroid on to its surface, would be as difficult as Burns' and Harwit's idea.

Instead, Sagan turned to what he calls 'biological amplification'—using a life-form to do the work, as in terraforming Venus. This time, he proposed using ground-growing plants to cover a small percentage of the ice-cap. Some Earth organisms are known that can survive Martian conditions, so it should be easy, thought Sagan, to breed suitable strains of plant that we could take to Mars.

These will have the added advantage of turning the plentiful but poisonous carbon dioxide in the atmosphere into breathable oxygen, so that men will be able to walk on the surface of Mars as they do on a spring day on Earth.

Nor should we forget our own Moon. It has no atmosphere of its own to speak of, because what little gas is released by the surface soon escapes into space. But in 1974, Richard Vondrak of Rice University, Texas, asked if this loss would continue if the atmosphere round the Moon were greatly increased. He came to the conclusion that the Moon could be made to hold a significant atmosphere.

Vondrak's calculations, published in the April 19th issue of *Nature*, show how the thin gases round the Moon are lost by heating and by the effect of the 'wind' of atomic particles blowing from the Sun. If the Moon's atmosphere were artificially increased above a certain amount then, according to Vondrak, a natural blanketing effect would prevent heated molecules from shooting off into space. Additionally,

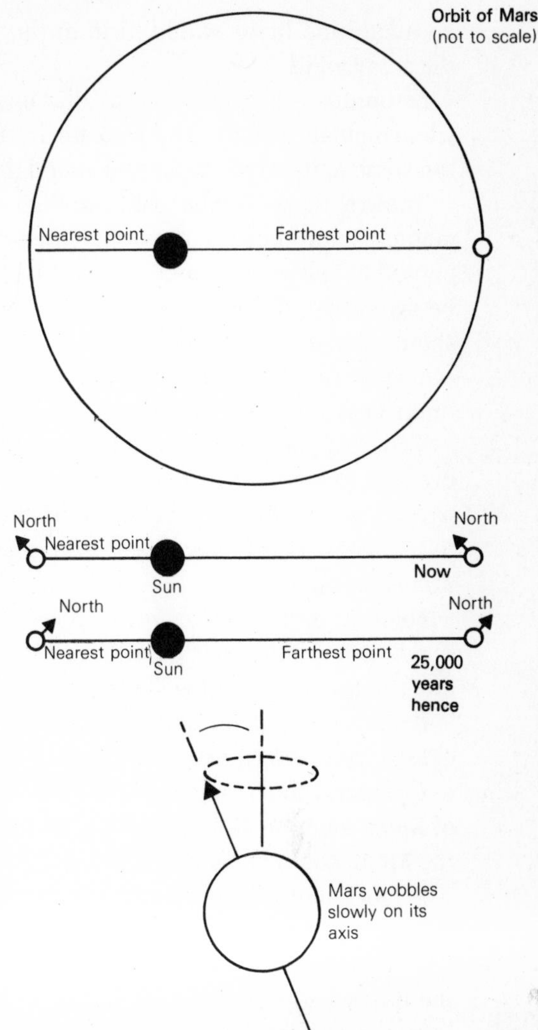

The wobble of Mars on its axis can affect the planet's climate in each hemisphere (see text).

an insulating layer would form at the top of the atmosphere to divert the solar wind.

Estimates place the current thin lunar atmosphere at about 10,000 kilogrammes weight. The rockets from each Apollo mission released the same amount of gas as this round the Moon.

Though these Apollo pollutants soon evaporate into space, a permanent lunar base and extensive exploration activities would be expected to release so much gas that the native lunar atmosphere would be swamped. Freeman Dyson has suggested that industrial processes should be moved to the Moon to prevent pollution of the Earth's atmosphere. But the result could be a lasting contamination of the lunar environment.

Vondrak goes on to consider that an artificial atmosphere could be created for the Moon by heating or vaporizing the lunar soil. In other words, we would be creating artificial volcanic eruptions—and it is the degassing of volcanoes that has given the Earth its present atmosphere.

Nuclear explosions might be necessary to melt enough of the rocks. If we found sufficient quantities of desirable materials on the Moon and planets, we could well want to use nuclear energy for mining.

Fortunately, as on Earth, the most abundant element in the Moon's crust is oxygen. Nearly two-thirds of the atoms in the Earth's crust are oxygen, and this trend seems continued in the relatively small number of Moon samples that we have gathered. If we can get this oxygen out, the Moon could be made into a comfortably habitable place.

A more immediately attainable goal, suggested by physicist Gerard O'Neill of Princeton, is to set up a community at one of the so-called Lagrangian points of the Moon's orbit, where the gravitational pulls of the Earth and Moon exactly balance. After O'Neill had canvassed this idea for 18 months, a public meeting was held at Princeton on May 10th, 1974, in which its full implications were discussed.

O'Neill believes that setting up such a space community is 'economically feasible, within the limits of the technology of this decade'. Initially it would support 10,000 people, who could then use free solar energy and the resources of the lunar surface to construct a still larger habitat. O'Neill foresees a progression leading to colonies of perhaps a million people 'possibly within 30 years from now. These communities would be as comfortable as the most desirable parts of the Earth, with natural sunshine, controlled weather, normal air, apparent gravity, and complete freedom from pollution,' he says. Such communities could grow to add new land area at a rate more rapid than the growth of the total human population.

O'Neill's proposed living-areas are on the inner surfaces of cylinders that spin to provide gravity. The insides of the cylinders would be divided into alternating strips of land and windows, with the effect of day and night created by peeling back or closing down reflecting mirrors from the windows. Light will always fall on the mirrors because each cylinder's axis points towards the Sun.

The composition and pressure of the air in the sealed cylinders is equal to that at sea-level on Earth, including clouds at heights typical of summer weather. Climate in the cylinder is controlled by the amount of sunshine that the mirrors allow in. No chemical fuels need be burnt for heat; and transport will be by bicycle and electric car. 'Therefore', says O'Neill, 'no smog.'

He also speculates upon some of the recreational opportunities—for instance, man-powered flight would be possible in the low gravity area near the cylinder's centre.

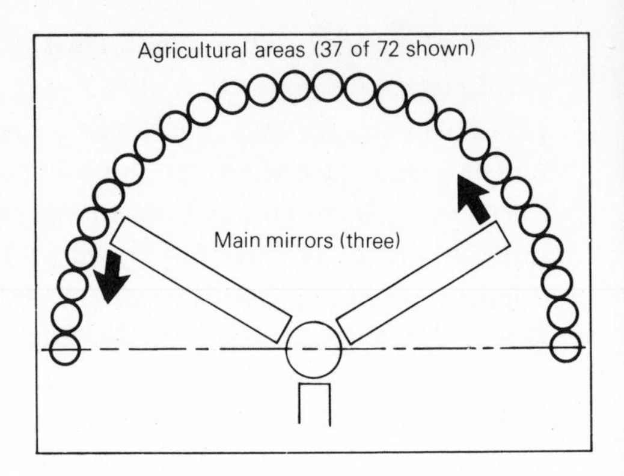

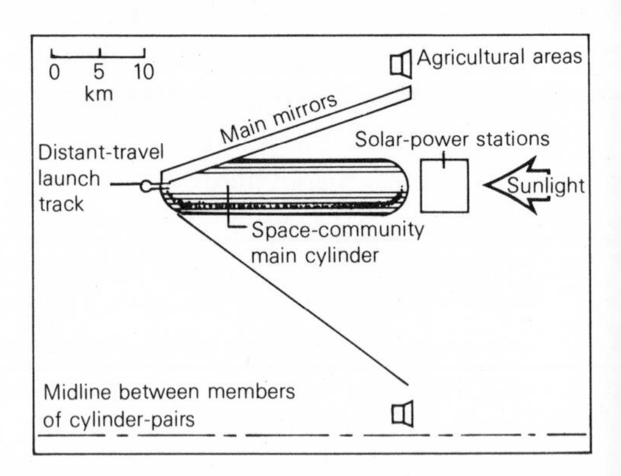

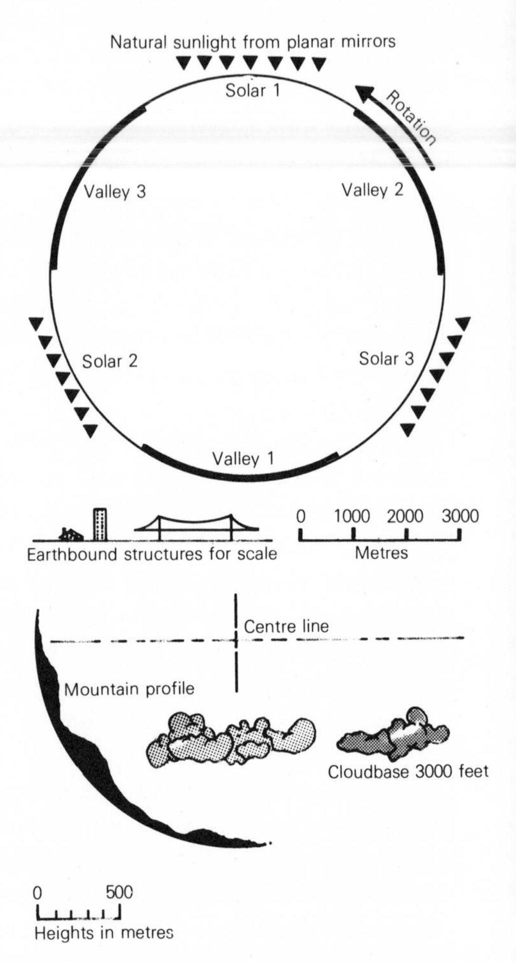

Space communities in cylinders, as proposed by Gerard O'Neill. Sunlight would be reflected into a giant cylinder by plane mirrors, which alternate with the living areas. Inside the cylinder would be natural air and even clouds; an artificial skyline of mountains might be constructed at the cylinder ends. Around each cylinder would be arranged small agricultural areas.

The first small cylinders would be about 1 kilometre long and have a radius of 100 metres. At their largest, the cylinders could stretch to 32 kilometres length and 3,200 metres radius. They would spin at rates ranging from once every 21 seconds to nearly 2 minutes.

O'Neill describes his view of this technological heaven: 'Bird and animal species that are endangered on Earth by agricultural and industrial chemical residues may find havens for growth in the space colonies, where insecticides are unnecessary, agricultural areas are physically separate from living areas, and industry has unlimited energy for recycling.'

Smaller cylinders arranged round the main living-cylinder would be the main food-growing areas. Each agricultural cylinder would have its climate controlled to suit the crop that it grew. Solar-power stations at the cylinders' ends would provide all energy needs, making the colonies totally self-sufficient. O'Neill stated emphatically in the September 1974 issue of *Physics Today*:

It is important to realize the enormous power of the space-colonization technique. If we begin to use it soon enough, and if we employ it wisely, at least five of the most serious problems now facing the world can be solved without recourse to repression: bringing every human being up to a living standard now enjoyed by only the most fortunate; protecting the biosphere from damage caused by transportation and industrial pollution; finding high-quality living space for a world population that is doubling every 35 years; finding clean, practical energy sources; preventing overload of Earth's heat balance.'

Because of its essential practicality (O'Neill presented engineering and economic considerations in his *Physics Today* article) this scheme is the most telling reply to those who maintain that success in space is no help to our pressing problems on Earth. It may become our first astro-engineering project.

As an extension of these ideas, we can see that the Solar System is going to be a pretty changeable place in the not-too-distant future. Scientists such as Carl Sagan are willing to talk about terraforming projects for the satellites of the giant planets, Jupiter and Saturn, or at least tapping the frozen gases on their surfaces for fuel.

Iosef Shklovsky, the Soviet astrophysicist, has been prominent in publicizing the advanced ideas of the Russian space theoretician Konstantin Tsiolkovsky, whom the Soviets term The Father of Astronautics.

Around the turn of the century Tsiolkovsky worked out the basic principles of spaceflight, when most people thought that even aeronautics was hokum. This pioneering Russian allowed his imagination to roam over the consequences of spaceflight, and in so doing he originated the concept of an energy-gathering shell round the Sun, which Freeman Dyson was to think up independently 65 years later.

In a book published in 1895, Tsiolkovsky foresaw that asteroids would become the horses of the Solar System—we would drive them round like small conveyances, using energy from the Sun. The technology was not then available, but Tsiolkovsky guessed that it would be developed—as indeed it has been, such as in the energy-giving solar cell, which translates sunlight into electricity.

Asteroids are likely to become a valuable source of metals, for there seems to be considerable iron and nickel in them, as evidenced by the asteroid fragments that fall to Earth as meteorites. An American space engineer, Dandridge Cole, discussing the value of the asteroids, estimated that the metals (including gold and platinum) in a three-mile-wide asteroid would be worth £25 million million. In addition, he spoke of harnessing an asteroid in Earth orbit as a natural space station.

No one will be much bothered if we scavenge the asteroids, or if we rearrange them in space to suit our astro-engineering needs.

But I suspect that astronomers as well as conservationists would be alarmed to find someone redistributing the mass of a planet in space. Once we upset the gravitational equilibrium of the Solar System, which has been attained over millions of years of subtle interactions between the planets' gravities, we shall risk altering the orbit of the Earth—with consequent alarming changes in our climate.

To chop up Jupiter and smear it into a Dyson shell near the Earth would radically disturb the balance of our space environment, for Jupiter weighs 2½ times as much as all the other planets put together. By the time that we are ready to move Jupiter, we should be able to keep our own planet's orbit in check, and possibly to engineer our own climate to suit ourselves. All this presupposes an unlikely concord of international views, but we can imagine that someone, somewhere, might do it.

Leaving Jupiter where it is, we can still use its massive store of hydrogen gas as fuel for nuclear reactors. This has been proposed by Iosef Shklovsky, who points out that there is enough hydrogen in Jupiter and the other giant planets made of gas in the Solar System to fuel a nuclear reaction that would release as much energy as a supernova.

If we used this energy relatively slowly—say, at the same rate as the Sun emits energy—the hydrogen store in the giant planets would last for approximately 300 million years. This, as Shklovsky sagely points out, is a time span '. . . most likely greater than the life of the technical civilization itself'.

Or, if we wanted to be really adventurous, we could borrow material from the Sun. A few per cent of the Sun's mass—far greater even than the raw material in the giant planets—would probably not be missed, yet it would provide an advanced civilization with enough energy in controlled thermo-nuclear reactions to last for thousands of millions of years.

Here, we are talking in terms of civilizations that correspond to Nikolai Kardashev's Type II—those with energy outputs equal to an entire star. These are the sort of civilizations from whom we would detect signals at galactic distances. We can see that in technological development they would be a few millennia ahead of ourselves.

Yet Kardashev also proposed a Type III civilization, able to tap energy resources equivalent to the power output of an entire galaxy. How would such a Herculean power release be generated? Astronomers can see that a single supernova gives out as much energy as an entire galaxy. If we can induce a star to the supernova condition, we shall be releasing energy of the order predicted by Kardashev for Type III civilizations.

Shklovsky suggests that a strong, concentrated laser beam 'might serve as the match which ignites the powder keg'.

And, as we saw in Chapter One, a supernova explosion makes the heavy elements from which planets form and which we mine for raw materials. So a civilization that has run short of raw materials might consider quarrying for them in the remains of a supernova. We would, however, be hard-put to distinguish an artificially triggered supernova from a natural one.

No matter how absurdly far-fetched these speculations seem, they may in fact be short-sighted compared with what waits to be found in the Universe. In centuries to come, we may well discover energy sources that are as unknown now as was nuclear energy a hundred years ago. Indeed, some scientists consider that certain cosmic objects, particularly quasars (these emit vast amounts of radiation and, according to some theories, they may represent an early stage in the evolution of galaxies) could be fired by new laws of physics. There may yet be power in the atom—or elsewhere—that is waiting to be tapped.

We shall certainly never be able to look back and say, 'All science was known in the twentieth century.'

In 1972 Freeman Dyson gave a lecture in which he put forward an idea that he had been toying with for a few years. Scanning the material of the Solar System, he had noted one salient fact that had escaped everyone's attention: comets are so numerous that they present a total surface area a thousand or ten thousand times that of the Earth.

Comets are small, ghostly bodies, but they exist between the planets and at the edge of the Solar System in vast numbers. 'Countless millions of comets are out there,' says Dyson, 'amply supplied with water, carbon and nitrogen, the basic constituents of living cells.'

And, from those considerations, he concludes that comets, not planets, are the major potential habitat of life in space. Dyson continues, 'They lack only two essential requirements for human settlement, namely warmth and air. Biological engineering will come to the rescue. We shall learn to grow trees on comets.'

Looking into the future, Dyson considers how we might implement this fantastic idea. Basically, he says, to make a tree grow in airless space by the light of a distant Sun, we need to redesign its leaves. We shall accomplish that, when we have the technique, by manipulating the genetic material of trees. A redesigned leaf must satisfy four requirements: it must have an adequate filter to prevent damage from ultraviolet radiation; it must not release water; it must let in visible light so that the life-giving process of photosynthesis can operate; and it must hold in the heat it produces, so that the leaf does not freeze.

Such a special leaf would allow a tree to grow on a comet as far away from the Sun as the orbits of Jupiter or Saturn. Beyond that, the leaf will need the still more sophisticated addition of a reflecting surface that focuses light on to the photosynthetic part like a small mirror. Perhaps thinking of the Venus flytrap, Dyson comments, 'Many existing plants possess structures more complicated than this.'

The rest of the tree has to be suitably modified to work in with this audacious scheme. The bark has to be a good insulator to stop the branches from freezing, and the oxygen and heat that the leaves produce during photosynthesis must be channelled down to the roots.

There, it will be released into the comet to create a habitable inner region where, Dyson foresees, men will live and take their ease among the tree trunks.

'How high can a tree on a comet grow?' he asks. Because gravity is so weak, there would be few shackles on the tree's development. A comet ten miles wide could produce trees hundreds of miles in length, collecting the light-energy of the Sun from an area thousands of times greater than the comet itself.

'Seen from far away, the comet will look like a small potato sprouting an immense growth of stems and foliage,' predicts Dyson. And what is more, the comet-dweller will be able to enjoy his arboreal existence in small, self-contained communities, far from the meddling influence of the powerful central governments that are the inevitable consequence of our thoughtless multiplication on Earth.

Gazing further into this fantastic vision of his crystal ball, Dyson

Living on a comet – an artist's impression of Freeman Dyson's imaginative idea. From the Ian Ridpath/David Hardy filmstrip *Exploring the Planets. (Hulton Educational Publications)*

touches on the possibility that the trees will make seeds that sail across the ocean of space to take root on comets as yet unvisited by man. 'Perhaps', he says, 'we shall start a wave of life which will spread from comet to comet without end until we have achieved the greening of the Galaxy.'

Although we see only those comets that are part of the Solar System, there may be many floating free in space of which we know nothing. If so, the distance between these 'habitable oases' would be much less than the several light years between stars—Dyson's own suggestion is one light day. And he points out that this would change the whole picture of interstellar travel.

Nevertheless, there is no evidence that any of the comets we have seen is or has been a habitation for extraterrestrial life—although we have seen only a tiny fraction of all comets.

Dyson developed his ideas from a book by the late J.D. Bernal called *The World, The Flesh, and The Devil**. Although written in 1929, this remains one of the most relevant prophesies of our future development, both on Earth and in space.

Bernal foresaw that man would leave his Earth and venture into space, possibly in a changed, mechanized form like cyborgs. 'Normal man is an evolutionary dead end,' he wrote. 'Mechanical man, apparently a break in organic evolution, is actually more in the true tradition of a further evolution.' Most of mankind, he felt, would prefer

* Jonathan Cape, London, 1970.

to remain behind on Earth, where its mechanized peers would regard it with detached interest as they explored space. Bernal suggested astro-engineering an asteroid to provide a space-station that would behave like 'an enormously complicated single-celled plant'. Dyson, by taking the comets, produced a novel method of fulfilling Bernal's vision.

But Dyson has gone further. He sees that, with the invention of programmable automatons, we now have the ability to make self-reproducing machines. These, Dyson says, would find their true realm in the regions of the Solar System that are inhospitable to man.

Machines need no water, so they can work happily on the bone-dry planets or among the asteroids. 'They will feed upon sunlight and rock, needing no other materials for their construction,' says Dyson. He imagines that they could use the raw material of the Solar System to perform major industrial tasks, possibly even bringing water from the outer planets to make the deserts of Mars bloom.

Taking a long look into the future, Dyson foresees that the Solar System will divide into two domains. The inner domain, near the Sun, is where light is abundant but water scarce. This is where the great machines would work as slaves of men, organized in giant bureaucracies. But in the outer domain, sunlight is scarce while water is plentiful. 'Here men will find once again the wilderness that they have lost on Earth,' says Dyson. The comet-dwellers will live in small communities, able to wander for ever on the open frontier that this planet no longer possesses.

The future of man in space could be idyllic indeed.

There are other technological alternatives for the future, and one of them is supplied by the weird natural phenomenon of the black hole.

Black holes are formed (so theorists think) when the central regions of a star are tightly packed by the explosion of its outer layers in a supernova. The gravity of this highly compressed object becomes so great that, in effect, its escape velocity is faster than the speed of light; therefore nothing can escape, not even light itself.

However, things can fall into a black hole and be swallowed up—including a spaceship. The fate of such a ship was examined by the Soviet astrophysicist Nikolai Kardashev in a presentation to the Byurakan conference on extraterrestrial intelligence. He pointed out that an object going into a black hole might emerge into a different universe. This raises the possibility—though not one to which many physicists would give much credence at present—that there is a rapid-transit system through black holes from one part of space and time to another.

But, more likely, black holes will turn out to be less of an open sesame to other worlds than a series of cosmic whirlpools that space navigators will want to avoid. Therefore, an advanced civilization might mark them with lighthouse beacons, which we could detect from Earth. Yet these are speculations, and we must now return to facts.

4

ORIGIN OF LIFE

We are made of the most abundant elements in the Universe: hydrogen, carbon, nitrogen and oxygen. The chemical composition of life is no mystery. The mystery is the assembly of those chemicals into the first living thing. What was the spark that first fired life? Biologists still puzzle over it. In the century since Darwin they have groped closer to an answer. But the ultimate truth of the matter still eludes them.

Darwin, in a phrase that echoes down the years, foresaw how life might have arisen 'in some warm little pond' where the right chemicals collected under the energizing light of the early Sun. From this snug little scene has grown the vision, fathered by J.B.S. Haldane, of a seething organic soup in the brimming seas of the primitive Earth.

In such a free-swimming environment, scientists think, the various chemicals of pre-life would be able to change partners until, by chance, suitable linkages were made that formed a substance which was more than an inert chemical. Washed on to the shore, and pegged out to dry in the warming Sun, such proto-life would be free to experiment for millions of years until, by the inevitability of chemical bonding, the first self-replicating organisms arose.

For food they would first gobble up the chemical stew in the surrounding water. When this feast wore thin, they would be forced to switch to photosynthesis—the use of sunlight's energy—to consume new food. Scientists see this as an important step for early life, for it gave it the freedom to survive anywhere on the globe.

Did a Divine hand guide this development of chemicals? The closer one looks, the farther back in time one is taken.

Laboratory experiments show that even dumb chemical assemblages are able to evolve by experiment as conditions change round them. Is this a repeat performance of the steps by which the first flickerings of life arose from that primeval soup?

There is no sign that the Universe has changed its laws over the history of the Earth, so we must suppose that what we see taking place now could also have occurred at our origin, as the right chemicals fitted together to give life.

It is no use asking *why* a particular chemical combination should give life. It is a result of the laws of the Universe, which equally govern the fact that gravity attracts us to the Earth rather than repels us, or that stars shine to give warmth. If the laws of the Universe were but slightly different, none of these things would happen. This fact has led many people to assume that the Universe must be custom-built for life.

But that is not necessarily so according to many astronomers, such as Dr Paul Davies of Cambridge University who reviewed the problem in the esteemed journal *Nature* in 1974.

A strong body of opinion asserts that there may be any number of universes, each with different laws. According to one widely accepted idea, the Universe continually blows up and collapses again, like a giant balloon.

After each cycle, which may take 100,000 million years, a new universe emerges like a phoenix from the ashes of the old, possibly with a completely new set of laws.

'Only very rarely does a cycle permit conscious organisms to evolve,' says Dr Davies.

We could not know of a cycle with different laws, because life would not then have formed. We are, by this account, a lucky by-product of nature's rule-book. So to the question 'Why are we here?' we must give the old soldiers' answer, 'We're here because we're here because we're here . . . !'

This new philosophy of life has come about as a result of the newest trends and developments in astronomy, which have all but thrown the need for a Divine hand entirely out of the Universe.

Yet the mystery of life's origin seemed so daunting that until recently scientists either failed to discuss it or, with a guarded eye on the Divine hand, preferred to attribute it to an agency from afar.

This latter approach was epitomized by the panspermia hypothesis of the Swedish chemist Svante Arrhenius, who proposed that the Earth had been seeded by spores from space. They would have drifted into our Solar System, expelled from another planet orbiting a distant star.

This would indeed be a remarkable odyssey: hibernation for millions of years, in an airless, frozen, radiation-soaked environment. It is an unpromising journey for an organism that is then supposed to germinate. Were it possible in practice—and it is not—it would merely displace the debate on the origin of life.

Independently in the 1920s, the Soviet biologist Alexander Oparin and the Englishman Haldane initiated the modern biochemical approach to the origin of life. In a way, they resurrected the spontaneous generation idea of life's origin that had been decapitated by the experiments of Louis Pasteur half a century before.

The charming, but factually hopeless, idea of spontaneous generation said that certain living things arose from the non-living, such as maggots from meat, and mice from straw. Pasteur's controlled experiments showed how contamination (and poor reasoning) could account for all the supposed examples of spontaneous generation. Inadvertently, Pasteur made the origin of life seem more of a puzzle than ever before, and put whole generations of biologists off the track.

Oparin and Haldane brought back the notion of the assemblage of life from random chemicals—but under rather specific conditions, when the Earth looked very different from the way it does today.

It is only since their time that any progress on the subject has been made; particularly in the past twenty years or so, since the decrypting of the genetic code which brought molecular biology of age.

The language of molecular biology is confusing. It contains many different terms, each describing a different type of chemical grouping. But there are only two main molecules on which we need to concentrate. They are the vital ingredients of all forms of life: nucleic acids and protein.

The nucleic acids, D.N.A. and R.N.A., guide the activities of a cell. The initials stand for deoxyribonucleic acid and ribonucleic acid respectively. They are the chemical blueprints for life.

Protein is an important constituent of cells, and it is also the structural matter of life; when we eat the remains of plant and animal life, we eat protein.

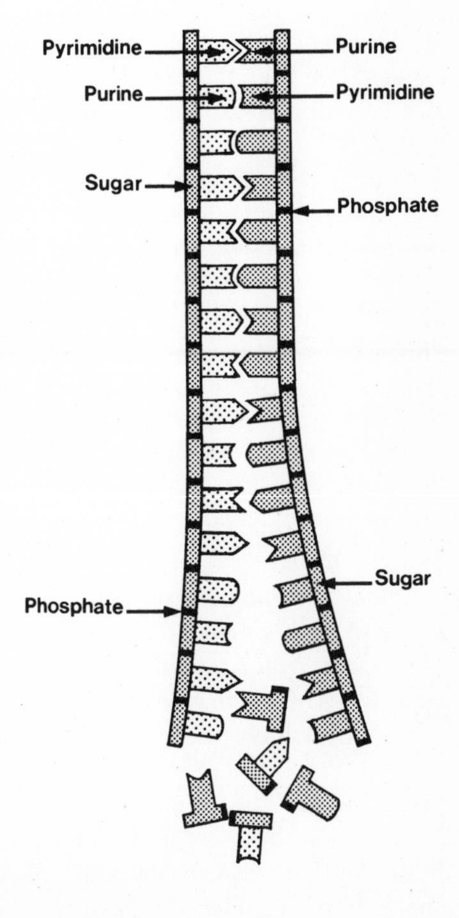

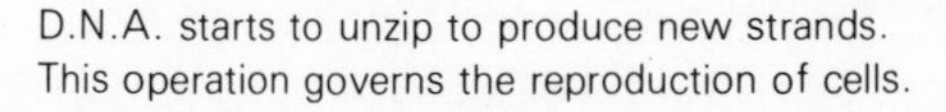
D.N.A. starts to unzip to produce new strands. This operation governs the reproduction of cells.

In a cell, which is the smallest biological unit in which life processes operate (tiny micro-organisms are often just single cells), the nucleic acid called D.N.A. unzips itself into two strands. It is D.N.A. that has the famous double-helix structure, which was first deduced by James Watson and Francis Crick in 1953. The links of each D.N.A. strand provide a chemical template for assembling protein. The other nucleic acid, R.N.A., acts as an intermediary in the assembly process.

The nucleic acids and proteins each have their own structural units, which are the same in every form of life on Earth, even though they may be arranged differently in different cells. It is these shared structural materials, plus the chemical pathways that can be traced out on the same map for all cells from microbes to man, that impress scientists in their belief in the unity of life's origin.

There are 8 structural units for nucleic acids, and 20 for proteins. The units slot together like the standard parts of a prefabricated building. Each structural unit is made of building-block atoms: hydrogen, carbon, nitrogen and oxygen, plus some others in small amounts. We saw earlier how these same atoms are created inside stars. Here is the key that links us all with the Universe at large.

To form into life, these common chemicals of space need a special order of assemblage.

Nature once performed that trick on Earth. And, incredible though it may seem, scientists have already gone part of the way to duplicating it.

They made their first step in one of the classic experiments of science, which was performed by the University of Chicago chemist Stanley Miller in 1953. In a flask he prepared a mixture of gases that were thought to have surrounded the primeval Earth: hydrogen, methane (carbon and hydrogen combined), ammonia (an alliance of nitrogen and hydrogen) and water vapour (hydrogen linked with oxygen). Through this blend he sent a high-voltage spark, in simulation of a lightning stroke passing through the atmosphere of the primitive Earth.

After a week of sparking, a rich mixture of new-formed compounds had rained down into the water at the bottom of Miller's flask. He had recreated the Earth's first organic soup.

The most exciting result was that among the compounds formed in Miller's experiment were amino acids, the structural units of protein. This was good news for the biochemists who were tracking down the origin of life, but it also presented a problem: several amino acids were formed that are not part of living things.

Why life should have preferred some amino acids to others is a puzzle that biologists have yet to solve. Also, they still need to find how all the 20 amino acids in protein can assemble naturally from their atomic building-blocks.

Since the Miller experiment, biochemists have mounted a baffling array of early-Earth simulations, adding to their brew small amounts of sulphur, which would have come from volcanoes on Earth, to form more types of amino acid.

Scientists have used ultraviolet light, which the Sun would have poured plentifully on to our planet, to form organic mixes from the primitive-atmosphere gases; they have fired energetic radiation at experimental containers of methane and ammonia, in imitation of storms from the Sun. And, by bursting plastic diaphragms, they have sent shock-waves through clouds of primordial-type gas to simulate the effect of meteorites plunging to Earth and thunderclaps rolling through the storm-filled skies of our planet.

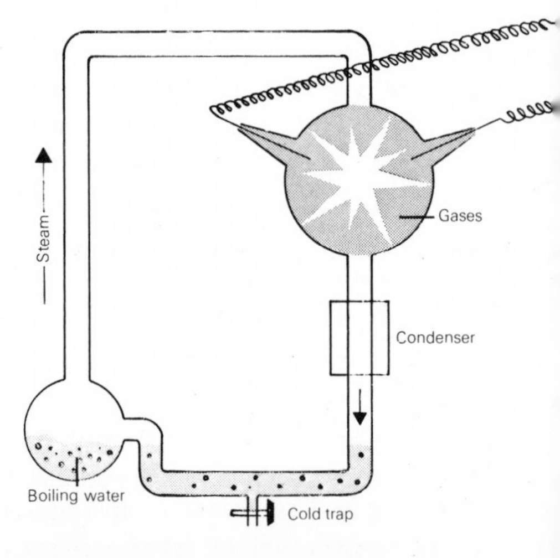

Diagram of Stanley Miller's apparatus that simulated conditions on the primeval Earth.

Amino acids have been formed by biochemist Leslie Orgel and his colleagues by freezing a liquid solution of primitive-atmosphere chemicals. The way that ice crystals line up—squeezing organic molecules into ranks between them—hints that chains of life's structural units could be built not in the baking conditions that most people have assumed but possibly in more frigid climes. But all these experiments have confirmed the fecundity of the early Earth's atmosphere for producing the first steps to life.

In particular, a group of Cornell University scientists engaged in these experiments concluded that shock-waves and ultraviolet light were 'of comparable importance in the prebiological synthesis of amino acids'.

These agencies could have produced organic molecules so plentifully in the environment of the primitive Earth that the first seas would have been swamped with them.

But ultraviolet light also slices molecules up, and the combination of its chopping and assembling would have produced an organic molecule solution in the ancient seas of a few per cent. This, as Leslie Orgel is fond of pointing out, is equivalent to the consistency of chicken soup—which supports life.

So there is no shortage of possible processes leading to the formation of the first soup of organic molecules on Earth. We can well imagine the scenes on our hot, damp planet as the clouds cleared and organic molecules rained gently into the seas below.

But an organic molecule—so called because it contains carbon and hydrogen atoms, which form the chemical backbone of life—is still far from living.

What is needed is a means of assembling the parts into the peculiar sequence necessary for life. Only when we can solve this puzzle can we begin to turn the origin of life from a mystery into a statistic.

The British science writer Nigel Calder has compared the implausibility of assembling the first proteins or nucleic acids by chance, with getting a jackpot on a fruit machine with hundreds of independently spinning wheels. What is needed, he points out, is some kind of 'hold' mechanism when the right wheel comes to rest in the right window.

The complexities of modern life-forms make this seem an impossible requirement. But the earliest molecules of life would have been much simpler. They need only have fitted together in a barely efficient way for life to have got its start. Then, the self-improvement programme of natural selection would have taken over.

The fittest molecules could not help but survive. They produced more and more efficient organisms, each capable of living and reproducing in a different habitat. We are the final outcome of D.N.A.'s mindless drive to replicate.

To solve the problem of how the molecules assembled to form life, scientists want to find ways in which the build-up of nucleic acids and proteins becomes inevitable.

The first hint of things to come in this direction was given in 1947 by J.D. Bernal, who suggested that organic molecules from the primeval soup could have further evolved on the surface of clays and rocks. The chemistry of clay-organic reactions remains a new area of study, but already some promising avenues have been opened.

Simulations of the hot, dry regions round primitive Earth's lakeshores have produced whole amino-acid chains. Other scientists have examined the role of extra chemicals to catalyse the marriage of each structural unit into a greater whole.

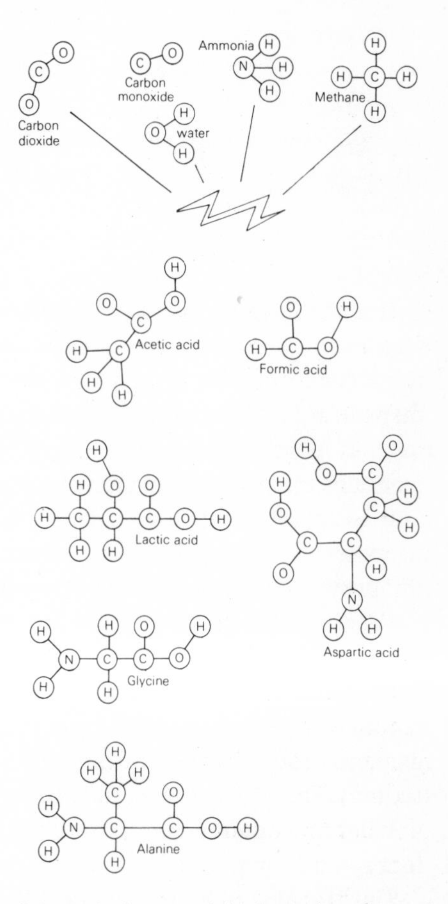

Schematic diagram of the formation of organic compounds in laboratory simulations of the early Earth.

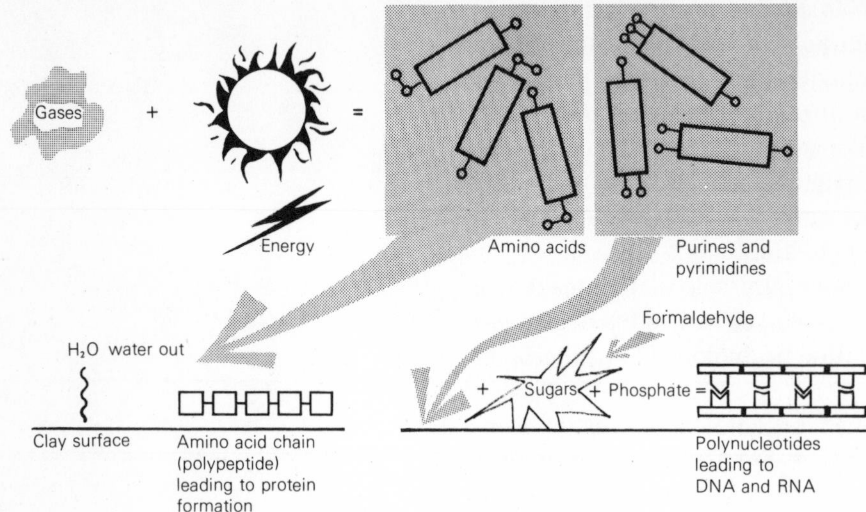

Possible sequences of events on the primitive Earth in the building up of life-forming molecules.

It is in this area that biochemists are likely to make the greatest new strides in understanding the mechanisms that lead to life. For the optimists, the clouds of mystery surrounding our origins are already dispersing. 'Laboratory experiments clearly demonstrate that a protein molecule could have been formed in the absence of life,' comments biochemist Cyril Ponnamperuma.

It would be foolish to assume that we now understand the exact processes by which the first big molecules of life were assembled, but the promise of doing so is certainly there.

Work by Leslie Orgel and his colleagues at the Salk Institute in California, published in 1973, throws considerable light on the chemical means by which the first protein and nucleic-acid chains could have been brought together, even to the point of a protein's assembly by a primitive nucleic acid. 'Most chemical evolutionists today think that complex nucleic acids and proteins evolved together, each acting upon the other to cause the formation of structures of ever-increasing complexity,' comments biochemist Peter Molton.

How far do we have to go before we can claim to have created life? Carl Sagan has this to say, 'The first life is a self-replicating polynucleotide [chain of nucleic acid structural units] in an ocean loaded with molecular building blocks — a polynucleotide which is able in a very crude way to determine the sequence of amino acids in neighbouring polypeptides [amino acid chains].'

Those sequences of amino acids then help on the reactions of other molecular assemblages round them.

'If you want to go further,' Sagan says, 'if you want to do work on the origin of the genetic code you are getting into a very elaborate molecular architecture which obviously took a long period of time to evolve.

'It may not be possible to do this in the laboratory because of the time involved.'

Yet Sagan remains optimistic that we shall find what he calls the 'trick' that set the early non-living molecules on the path to elaborately reproducing life. Will the same trick be worked on other planets? It depends on how rare a trick it is, he replies. 'So far there is nothing that is not common to other planets in the steps to the origin of life that have been done in the laboratory.'

Sagan does not identify the first cell with the first life. That is a more complex step that would have come later, once the self-replicating molecules were well established on Earth. But it is a step that we still need to explain in order to understand our own evolution.

Even here, there are the glimmerings of an answer.

Prominent among researchers in this direction has been Sidney Fox of the University of Miami. He has found that heat can turn amino acids into chains whose properties, with some differences, 'resemble those of proteins.'

Fox believes that such *protenoids*, as he calls them, formed on the dry land around hot springs on the ancient Earth. There is evidence that in similar regions today the Earth is capable of building up such organic compounds. Fox finds that, under these conditions, most of the amino acids in protein assemble themselves, with little trace of the non-protein amino acids. After a fall of rain, these protenoid chains would be washed into solution. With what result? They would spontaneously produce structures that look and behave remarkably like early cells.

Fox, in his experiments, has demonstrated that protenoids form into what he calls *microspheres* when they are put into water. The microspheres have double walls—long thought to be a property only of modern cells—and they have even been seen to split, like the division of a living cell. Even expert audiences have mistaken slides of these microspheres for bacteria.

Although the first nucleic acids may also have arisen under similar conditions, they were clearly not essential for the assembly of the first amino acids.

In an issue of *Chemical and Engineering News* for June 1970, Fox published photographs to show how closely some of his protenoid spheres resemble the spores of early life discovered in ancient Earth rocks by Elso Barghoorn of Harvard.

'The experiments suggest', says Fox, 'that evolution of molecular complexity was capable of occurring from simple beginnings very rapidly—in days or less.'

Yet, as we have seen, these protein spheres of Fox have none of the genetic material, the D.N.A., to guide their reproduction or the assemblage of new living tissue. They are genetically dumb. But so, it seems, may have been the first cells on Earth.

Another angle on the origin of organized life on Earth was given by an American team late in 1971, in a paper in the magazine *Science* startlingly titled 'The Primordial Oil Slick.' Their calculations showed that the methane gas in the early Earth's atmosphere would have been joined together by the Sun's ultraviolet light to make a complex hydrocarbon liquid—oil. This would have formed over a period of a million or so years, and coated the surface of the primeval sea with an oil slick between 3 and 30 feet thick.

Therefore, the dilute pre-biological soup would have been covered by a rich hydrocarbon layer with, the authors state, 'some rather intriguing properties'. For one thing, it puts the possible origin of the first cell-like structures firmly back into the ocean again, rather than in the fuming mudbanks of the ancient shores.

Writing in the magazine *Spaceflight* in 1972, Dr Peter Molton of the University of Maryland described how the oil would act as a solvent for organic compounds from the water below, extracting them for further activity in the pre-biotic pageant.

Again, clay particles come into the act. They would form a kind of nucleus which collected organic compounds. The oil would provide a barrier between these and the ocean; a valuable barrier, for one of the problems in the first steps of life is that water must be removed before the long molecule chains can join up.

This has been possible on dried clay—but not in the soupy ocean.

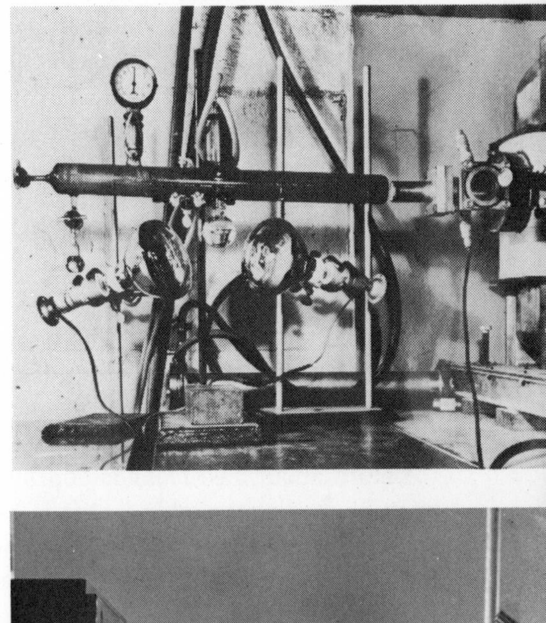

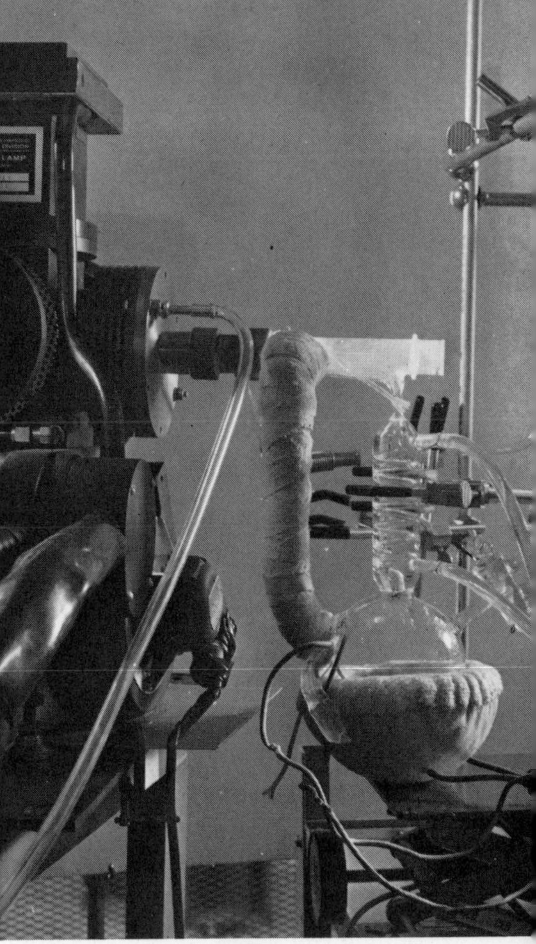

Experiments to produce organic compounds from primitive atmosphere gases. Above, radiation is shot in from a particle accelerator at the right, and organic compounds fall into the small round flask under the long tube. Below, ultra-violet light is injected into a flask from the lamp at the left to create organic compounds. (*Courtesy Cyril Ponnamperuma*)

Since oil and water don't mix, Dr Molton points to this as a possible mechanism for allowing the build-up of complex organic molecules in solution.

In all this work, scientists are backing the simplest assumption: that cell-like structures of protein came first, and later picked up the other chemicals that make the modern cell so complex.

Yet, once we have life, what are the chances that it will develop into more complex forms? Francis Crick, Nobel prize-winner for his part with James Watson in unravelling the spiral-staircase structure of D.N.A., maintains that once a certain degree of cellular organization is present 'there is bound to be an evolution to organisms with many cells and to many different organisms of increasing degrees of complexity as time goes on.'

The truth about those first steps in life lies nearly 4,000 million years back in the Earth's history. Or millions of miles away on another planet.

Improved understanding of the beginnings of life has led many scientists to the belief that the emergence of living creatures on Earth was chemically inevitable—Carl Sagan terms it a 'forced process'. This idea is supported by geological evidence.

Before about 3,900 million years ago, conditions on Earth were too unstable and changeable for life to have gained a steady foothold. The melting of the Moon's surface attested to by the Apollo rocks shows that there was until that date a great release of heat from the newly formed members of the Solar System. A massive bombardment was then still in progress, as the planets swept up the debris left over from their formation.

But after that time things would have become stable enough for life to have sprung forth as soon as it was ready.

According to geological evidence published in 1972, there are signs of photosynthetic life as far back as 3,300 million years ago. This is a remarkably advanced state of life for so early in our planet's history.

The evidence comes from analyses of the carbon content of rocks, which is affected by the presence of carbon-based life. Detectable changes in the carbon were found at the 3,300 million-year mark, which suggested to the experimenters that simple organisms had passed the threshold to photosynthetic life at that time.

Such a date is supported by more direct evidence: what are thought to be the remains of blue-green algae have been seen in rocks well beyond 3,000 million years in age. Biologists can recognize these photosynthetic organisms because some forms of them have survived virtually unaltered to the present day.

Then, in 1974, at a conference on alien life organized by the British Royal Society, Leicester University geologist Professor P.C. Sylvester-Bradley argued persuasively that indications of life go back farther still—to the oldest dateable rocks, about 3,800 million years ago.

From this, University of Edinburgh geneticist Conrad Waddington was willing to speculate on a period of 'explosive' evolution once the first chemical precursors of life had formed. He believed that 100 million years was more than sufficient for life to have progressed from that stage to the origin of photosynthesis, as practised by the ubiquitous blue-green algae.

Again, nature has surprised us with her ready ability to experiment, showing that the optimistic scientists with their laboratory flasks of early-Earth atmosphere may have been far closer to duplicating the first steps towards living beings than even they had dared hope.

And, as the origin of life on Earth is pinned down to a short and ever-earlier period in the Earth's history, our optimism increases that it could also have happened elsewhere.

'The Earth becomes a model laboratory for what may have happened on innumerable occasions in other solar systems,' says University of Maryland biochemist Cyril Ponnamperuma, who has been responsible for outlining many of the chemical steps towards life.

Spectacular evidence of the natural origin of organic molecules elsewhere in space came in 1969 when a meteorite landed near Murchison, Australia, with an on-board complement of amino acids similar to those created in the laboratory.

Meteorites are rocks from space that occasionally crash to Earth. They probably all originate within the Solar System, and must have been formed at about the same time as the planets. So when we examine a meteorite, we are in effect looking at the primitive material from which our own Earth formed. Meteorites still retain the stamp of the conditions round the Sun about 4,500 million years ago. This includes traces of the chemical fog that gave us our first atmosphere. Some groups of stony meteorites contain a certain amount of carbon. They are called carbonaceous chondrites, from the fact that they also contain small lumps or *chondrules* made of minerals.

False alarms about the organic content of carbonaceous chondrites had been raised before. One famous example was the Orgueil meteorite that fell in France in 1864. It was claimed that analyses of the meteorite made a hundred years later showed the existence of complex organic compounds and even tiny cell-like shapes, looking like spores.

But, in the time since its fall, the meteorite had plenty of time to absorb terrestrial contamination, and the most spectacular of the claims about Orgueil and its fellow meteorites have since been discredited. 'What appears to be the pitter-patter of heavenly feet is probably instead the print of an earthly thumb,' said one sceptical chemist.

By contrast, a newly fallen meteorite has no time to become contaminated. The Murchison meteorite proved to be one of the group of carbon-containing meteorites in which researchers had previously believed they had found organic molecules from space.

Fortunately, it fell at a time when space scientists in the United States were developing ultra-sensitive and spotlessly clean analytical laboratories for studies of the first Apollo Moon rocks. They were able to try out these new techniques on the Murchison meteorite.

At the Ames Research Center in California, a team of scientists found a number of amino acids in the Murchison meteorite, all apparently of extraterrestrial origin.

Organic molecules of biological origin—that is, made by living things on Earth—have a certain odd property that distinguishes them from other molecules. Molecules with an identical chemical composition can be joined in a left-handed or right-handed way. They are mirror images of each other. The molecules of life on Earth choose a particular left- or right-handed form. The form can differ among different types of molecules. Amino acids in living things, for example, are left-handed.

This specific 'handedness' is a curious property of life-giving molecules on Earth which presents a puzzle for biochemists. Organic molecules synthesized in the laboratory experiments have roughly equal numbers of left- and right-handed molecules. So also should the first organic molecules that were made on the primitive Earth.

Why did one particular type come to dominate?

Professor Cyril Ponnamperuma with a sample of the Murchison meteorite, in which he has helped find traces of many organic compounds.

Left or right? Carl Sagan ponders the mirror-image nature of life. (*Ian Ridpath*)

Probably because the numbers of left- and right-handed forms were not *exactly* equal. Computer simulations show that, even if the difference is one in a million, one form will rapidly take control.

The handedness of nature—which also underlines the unity of life on Earth—is an excellent way of distinguishing terrestrial contamination from organic molecules formed in space.

As the Ames Research Center team found, the Murchison meteorite contained equal numbers of left- and right-handed forms of each amino acid—exactly what would be expected if they were synthesized non-biologically in space. For the research team's head, Dr Cyril Ponnamperuma, the discovery was 'probably conclusive proof' that the same molecule-forming processes operated in space as those he had simulated in the laboratory.

Shortly afterwards, Ponnamperuma and his colleagues examined a specimen of the Murray meteorite, which had fallen in Kentucky on September 20th, 1950. They found the same batch of organic compounds as in the Murchison meteorite. The great chemical similarity between these two meteorites, and the nearly coincidental time of the year on which they fell, suggests that they were fragments of the same larger body, orbiting the Sun in a path that crosses the Earth's orbit. Alternatively, such organic compounds may have been widely distributed in the Solar System at the time the meteorites were formed.

The team at Ames Research Center then looked at samples of the controversial Orgueil meteorite, and were able to distinguish organic compounds inherent in the meteorite from those caused by terrestrial contamination. The previous researchers had been unable to make this distinction.

But amino acids were not the only organic molecules to be found in the Murchison and Murray meteorites. Two other molecules, of the type known as *pyrimidines*, were also present.

Pyrimidines are simple six-sided molecules. Two types are represented in D.N.A.; one of these, along with a third type, appears in R.N.A. Several pyrimidines were found in the Murchison meteorite, and similar molecules were also present in samples from the Murray, Orgueil and Allende meteorites. None of these pyrimidines occurs in living things, and can only have been formed non-biologically beyond the Earth.

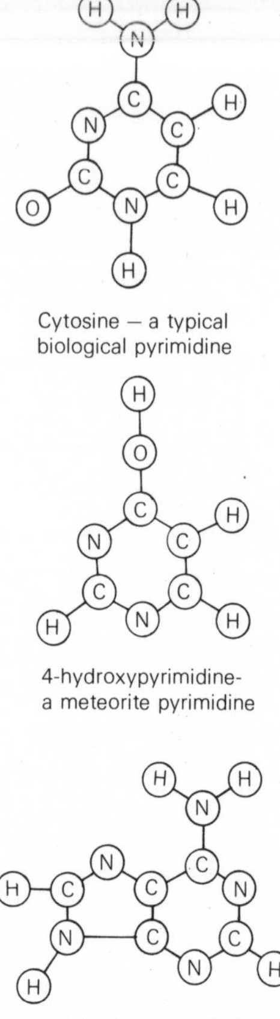

The chemical structure of a purine and two pyrimidines.

Comet Kohoutek as many people saw it in January 1974. (*Royal Greenwich Observatory*)

Why those pyrimidines were found, and not the ones in living things, is a puzzle to biologists. Yet the biologists have equal trouble synthesizing the pyrimidines of life in their laboratory experiments. They have had to try making them by sticking together the less complex molecules from the pre-biological organic soup.

Another important group of structural units in the nucleic acids are the *purines*. These consist of a five-sided molecule bonded to a six-sided ring. One such purine, called adenine, was made in the laboratory in 1961. Adenine is a combination of several atoms of hydrogen, carbon and nitrogen (symbols H, C, and N). A simple combination of these atoms is hydrogen cyanide (HCN), and so it seems that hydrogen cyanide in the Universe is a precursor of adenine.

Hydrogen cyanide is made plentifully in the experiments with primitive atmospheric gases, and we can see that hydrogen cyanide is abundant in the Universe. There is, for instance, plenty of hydrogen cyanide among the gases of comets, as exemplified by the famous Comet Kohoutek of 1973-4.

In a *Nature* paper published in 1974, Soviet space scientist Lev Mukhin concluded from his investigations of the volcanic areas of the Kuriles and Kamchatka, that ten tons of hydrogen cyanide could be given out by a single eruption. 'Volcanism must have played a significant role in the synthesis of organic compounds,' he said.

Researchers into the origin of life are encouraged by the fact that many of the other simple organic compounds that they have produced in the laboratory are now known to exist in comets. And the ghostly comets, even more than meteorites, provide a living-fossil record of the material that once swarmed round the Sun.

Adenine has been found to be a product of the experiments that sent streams of energetic radiation into the mix of primitive atmospheric gases. It seems that adenine could have been easily formed on Earth by radiation from the Sun.

But there are other purines that have not so far been made in experiments simulating early-Earth conditions. This is another of the avenues that further research is likely to open up in the near future.

Other structural units of nucleic acids are the sugars. As yet, none of these has been directly produced in the primitive-Earth experiments. But a common organic substance, called formaldehyde, is a simpler version of such sugars. Organic chemists believe that formaldehyde molecules can combine to give the sugars found in D.N.A. and R.N.A.

Formaldehyde is a combination of hydrogen, carbon and oxygen (formula H_2CO). From this simple arrangement, we might expect it to be formed easily from those same gases that are thought to have been plentiful round the primitive Earth, and laboratory experiments show that this is indeed so.

The Allende meteorite, a carbonaceous chondrite which fell in Mexico in 1969, also brought to Earth evidence for formaldehyde in space. A team from the U.S. Geological Survey and the Smithsonian Institution in Washington analysed the Allende meteorite in 1972 and reported the existence of formaldehyde in it.

Another class of organic molecules has also been found in the carbon-containing meteorites—though these are not part of proteins or nucleic acid. They are the fatty acids, which are made of chains of carbon, hydrogen and oxygen atoms. Seventeen such compounds were found late in 1973 in the Murchison and Murray meteorites by researchers at Ames. The compounds include acetic acid and propanoic acid, both of which are formed in laboratory experiments on organic chemical synthesis.

Slightly more complex substances, called dicarboxylic acids, were found in the bountiful Murchison meteorite in 1974. These substances are important because they take part in the metabolic cycles of living things.

Simultaneously with this discovery, an Ames Research Center team found that they could make those same substances in laboratory electric-discharge experiments.

Most staggering of all to the scientific community was a report from Moscow in 1972 of a double-spiral structure, like the helix form of D.N.A., found in the Mighei meteorite which landed in the Ukraine in 1889.

The finding, reported by Alexander Vinogradov and Gennady Vdovykin of the Vernadsky Institute of Geochemistry and Analytical Chemistry, has particularly far-ranging implications. It shows that large molecules, more complex than any before known, could have been formed in the gases round the early Sun, in addition to the more hospitable regions on Earth.

Although the Soviet scientists do not claim that the polynucleotide in the Mighei meteorite is a direct precursor of D.N.A., it is nevertheless an impressive analogue, particularly as it is found in a carbonaceous chondrite in which the usual complement of amino acids and other organic molecules are also present.

The spiral polynucleotide that Vinogradov and Vdovykin found was symmetrical in structure, unlike biological D.N.A., and was evidently formed on the surface of dust grains, and at low temperature, from simple atoms in the gas cloud round the Sun. The Sun's radiation would have provided the energy to stick the substances together, in confirmation of the building processes already deduced for organic molecules in the clouds of the primitive Earth.

Moon rocks have also been studied for possible organic content. Traces of amino acids were found in the Apollo 11 and 12 rocks, and in 1973 were confirmed in Apollo 14 samples. The amino acids found were similar to those in meteorites, and were not due to terrestrial contamination. They could have been formed as the Solar System passed through a cloud of interstellar matter, because samples from below the Moon's surface show fewer or no amino acids. Other

A scientist examines Apollo Moon rocks under ultra-sterile conditions. The delicate analytical techniques developed for finding organic material in Moon rocks were also used to find complex molecules in meteorites. (NASA)

suggestions are that the amino acids form on the Moon's surface from the solar wind or are remnants of organic compounds from meteorites that struck the Moon.

By a strange twist of scientific fate, these developments have led to a revival, in modified form, of the panspermia hypothesis of life's origin on Earth.

After their publication in March 1972 of the formaldehyde content of the Allende meteorite, the American research team who made the finding said that such discoveries of organic material in meteorites 'give us cause to reconsider some of the broader theories regarding the origin of life'.

They calculated that a total of over 100 million tons of formaldehyde and amino acids could have been delivered to Earth by meteorites of all sizes before the first life arose on this planet. (Extra material might be added by encounters with comets, but the authors did not take this into account.)

From their calculations they concluded in their paper in the scientific journal *Nature* that:

> Certain compounds that exist in space or may be formed on meteoritic surfaces or within meteorites can be distributed by those meteorites and, on landing on a friendly body, may well serve as the precursors of life.
>
> Proteins and related substances are to be expected in the case of amino acids; in the case of formaldehyde, carbohydrates are the anticipated product.

The following year, biochemists Leslie Orgel and Francis Crick took a different approach in a paper in the planetary magazine *Icarus*. Acknowledging that accidental panspermia is effectively a non-starter, they instead proposed that the Earth could have been deliberately seeded by organisms from another civilization, reflecting the ideas on interstellar contact presented in Chapter Nine.

They imagined that the civilization in question might be facing the death of its star, or that it could have other, possibly more megalomaniac, reasons for transmitting its organisms to another planet.

Crick and Orgel examined the technique of sending life forms over interstellar distances, and concluded that a rocket full of micro-organisms would be a plausible way of effecting the transfer. While not suggesting that this was the only way that life could have arisen on Earth, they believed that present knowledge is unable to rule out conclusively such an idea.

Crick and Orgel adduced two 'weak arguments' in support of directed panspermia. One is the apparent importance of the comparatively rare substance molybdenum in terrestrial biochemistry—whereas the more common elements chromium and nickel are, they said, 'relatively unimportant'. The chemical composition of living organisms should closely reflect the composition of their environment. In another planetary system, however, the chemical elements might come in different relative amounts, and the authors therefore suggested a search for molybdenum-rich stars.

They also referred to the fact that all organisms on Earth have a similar chemical basis for reproduction—the genetic code. 'It is a little surprising that organisms with somewhat different codes do not coexist,' they commented. Their solution: that life originated from a single micro-organism from another planet. 'Present-day organisms should be carefully scrutinized to see if they still bear any vestigial traces of extraterrestrial origin,' they suggested.

Ensuing criticisms suggested that Crick and Orgel had got their figures wrong in estimating the relative abundances of molybdenum, chromium and nickel. In particular, one critic noted, molybdenum is most abundant in sea water, where life is expected to have arisen on Earth.

Undaunted, Orgel replied that it has yet to be shown that the molybdenum content of sea-water has always been the same. And there the matter rests.

Perhaps the most staggering new avenue in the search for life's origins was opened a few years ago by the work of radio astronomers.

They found organic molecules in the gas-and-dust clouds of space—a discovery which showed that the steps leading to the build-up of the structural units of life can operate even in the cold, airless conditions between stars.

The first molecules to be detected in space were simple linkages of two atoms. At that time, no other combinations were expected.

But now, around 30 molecules of increasing complexity are known, and a major body of literature has been published on the subject. It is even possible that only the simplest molecules are at present being detected, and that a whole host of complex organic compounds, rivalling or even exceeding those in meteorites, remains to be identified.

Such molecules can be detected in space because they give out or absorb radiation at particular wavelengths. These wavelengths lie in the microwave part of the spectrum, which is not visible, so the investigations have to be done with radio telescopes.

Each molecule is identified by its own spectral fingerprint, like the

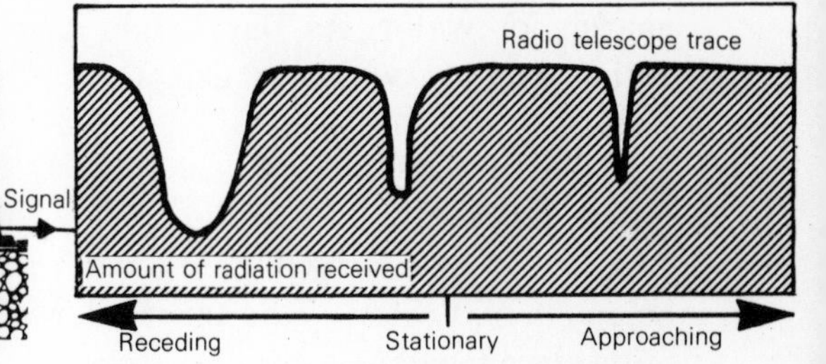

Absorption of radiation coming from the centre of our Galaxy by three gas clouds containing formaldehyde. The motion of two of the clouds is discernable by the differing positions of maximum absorption on the signal received by the radio telescope.

pattern of lines in the optical spectrum that reveals the presence of different gases in the Sun and stars.

Spectroscopy is a major method of analysis for scientists, and is pursued at all wavelengths—for the molecules of the Universe are not so kind as to trade all their parcels of energy in the visible part of the spectrum. This particular branch of spectroscopy is so new that the theory behind it is still not fully understood. Much of the current uncertainty about detecting molecules between the stars arises from the fact that scientists are still unsure what radio wavelengths to tune in to. For instance, amino acids would be difficult to find because their microwave fingerprint has not yet been measured in the laboratory.

Exactly how the molecules emit or absorb radiation at microwave frequencies is somewhat technical, but the general idea is this. The molecules in the dark, cool clouds of space continuously tumble end over end. Occasionally, the rate of tumbling may change, perhaps as a result of a collision with another molecule. When that happens, the molecule gives out radiation, usually at a wavelength of about one centimetre. If it is hit by a beam of radiation of that same wavelength, the molecule will absorb it.

Another way for the molecules to emit signals is by their natural vibration. The atoms in the molecule oscillate back and forth, and again the molecule gives out centimetre-wavelength radiation. Also, they can absorb radiation of the right wavelength.

Because these changes occur at low temperatures, the emitted or absorbed energy is correspondingly low. The spectral fingerprint is therefore only detectable in the radio region of the spectrum.

In hot gas, by contrast, there are high-energy changes in the orbits of the electrons surrounding each atom. These produce lines in the visible portion of the spectrum, as happens in stars. But the interstellar gas is too cold for this particular emission mechanism to work.

The nature of the signals emitted by the interstellar molecules depends on the density and temperature of the cloud they are in. These signals therefore act like interstellar probes for astronomers, telling them about the conditions between the stars.

Not all the molecules found in space are organic—that is, they do not all contain carbon and hydrogen—but they all contain atomic combinations that are familiar to chemists on Earth.

One of the early exciting discoveries of molecules in space was the emission of intense radiation from a hydrogen-oxygen combination known as hydroxyl. The formula for this molecule is OH. An extra hydrogen atom would produce water, H_2O.

The hydroxyl emission lines were discovered in 1965, and in 1968 two American groups made plans to search for more complex molecules in space.

At the University of California at Berkeley, a group led by Charles Townes built a special 20-foot radio telescope to search for ammonia in space, while at the National Radio Astronomy Observatory, Lewis Snyder and David Buhl prepared the 140-foot Green Bank telescope to search for water. As David Buhl describes it, the competition was rather exciting, with most astronomers betting that both groups would lose.

In the event, both were spectacularly successful.

The Berkeley team took the lead by getting their equipment working first. In the clouds towards the centre of the Galaxy they found both ammonia and water—the emission frequencies of both molecules are quite close together.

Ammonia (NH_3) has since been found to be quite common in the large dust-and-gas clouds of the Galaxy, and water has proved to be a particularly revealing substance; the conditions for water-emission signals occur only in small, dense clouds where stars are forming.

Having been beaten in the race to find the first complex molecules in space, the Green Bank team of Snyder and Buhl turned their attention to formaldehyde (H_2CO) which seemed a likely candidate for discovery once the existence of ammonia and water was established.

To the team's joy, in 1969 it became apparent that formaldehyde is a particularly ubiquitous molecule in our Galaxy. Now, formaldehyde is also known in galaxies outside our own.

In their Earth-bound laboratories, the organic chemists were not slow to react to these astounding discoveries that threw a penetrating new light on the quest for the origin of life.

In 1970 at their University of Miami laboratory, biochemists Sidney Fox and Charles Windsor heated together formaldehyde and ammonia—and got a number of amino acids, predominantly glycine and alanine. These are the amino acids that were produced most readily in Stanley Miller's first experiments, and which are also most common among the amino acids found in the Apollo Moon samples. Commenting on their technique, Fox and Windsor wrote, 'Heat is generally less destructive than high-energy radiation.' Their work showed, they added, that amino acids can be built from precursors in contemporary interstellar matter, rather than postulated primitive atmospheres.

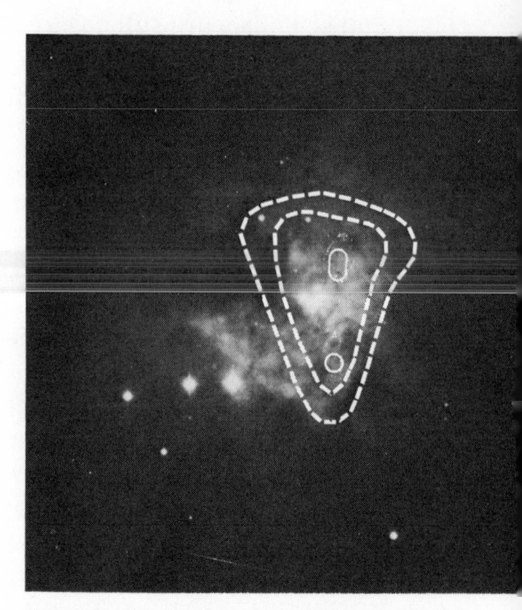

Contours of formaldehyde emission from the central part of the Orion nebula. A cloud of hydrogen cyanide that is twice as big has also been detected. The top peak of the formaldehyde emission is thought to coincide with the site of a forming star. (*Lick Observatory photograph; data from Barry E. Turner*)

Referring to Fox's earlier work on producing cell-like spheres from protein chains, the researchers commented that the origin of a sufficient number of amino acids can now be accounted for, to place the total sequence from simple compounds to microspheres 'on a relatively factual basis'.

In the wake of these initial discoveries, radio astronomers found a whole host of other complex molecules, both organic and otherwise, in the dust-and-gas clouds between the stars.

Methane has never been detected, because its molecular structure does not allow it to emit a suitable signal but, by analogy with similar chemical structures that have been found, it almost certainly exists.

Radio astronomers are therefore confident that the main ingredients used in the laboratory mixtures to simulate the primitive Earth atmosphere—ammonia, methane and water—are to be found in interstellar matter.

Astronomers still debate how complex molecules are built up in space, but it seems likely that they form on the surfaces of so-called dust grains. These are particles the size of those in smoke to which individual atoms or atomic linkages can stick, building up a more complex molecule as a result. This can be propelled into space by the chance impact of high-energy radiation, such as a cosmic ray.

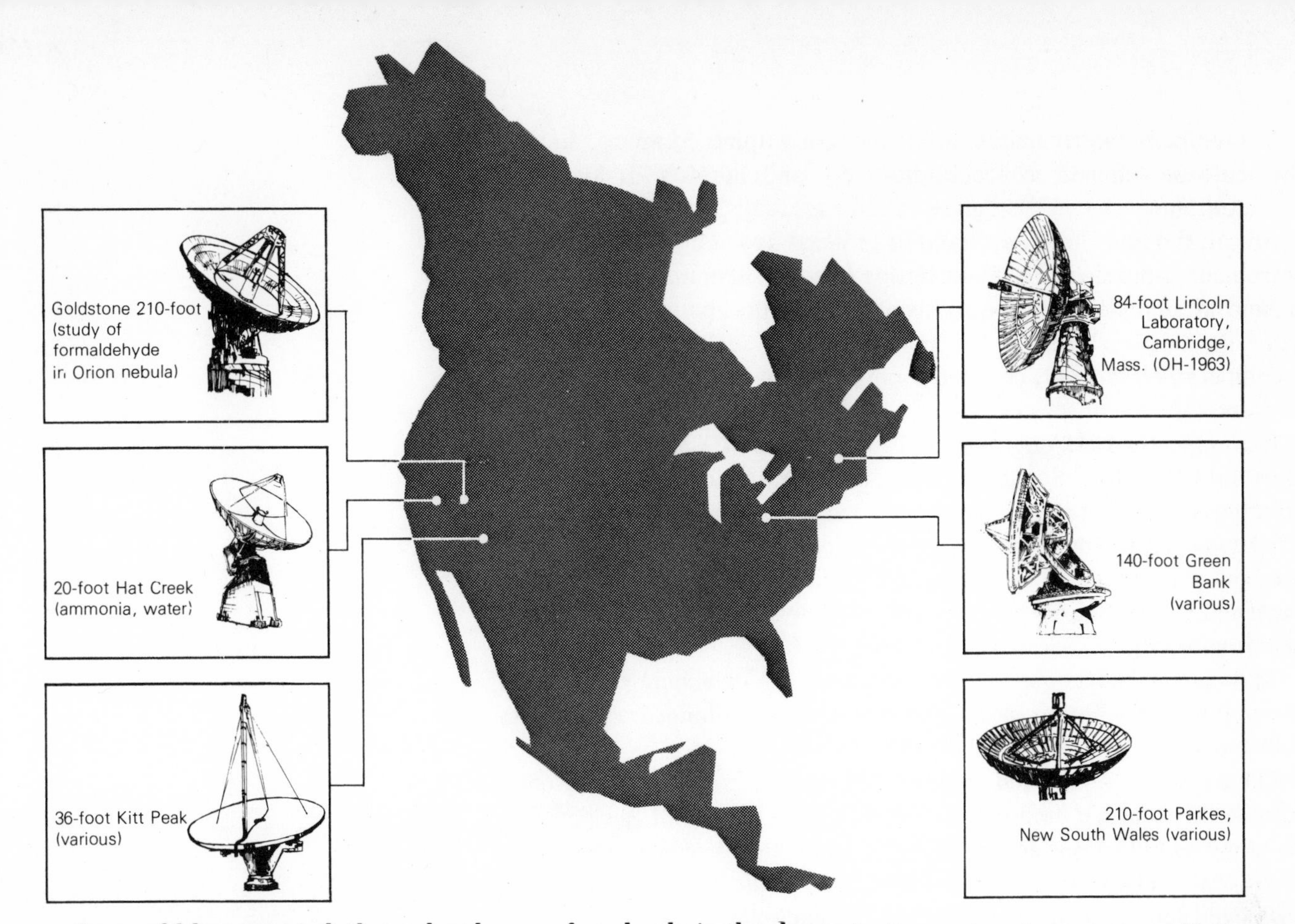

Some major radio telescopes used in the search for interstellar molecules. (NASA, University of California, NRAO, U.S. Air Force, Australian News and Information Service)

As would be expected, the molecules are found only in the densest clouds between the stars, because only there do sufficient dust grains and gas particles collect to build up a significant cosmic chemistry. The density of the clouds, in addition, provides a shield against ultraviolet radiation which would otherwise break up the molecules in a very short time.

One of the most significant early discoveries of molecules, made in 1970, was that of cyanoacetylene (formula HC_3N). This is one of the chemicals that laboratory scientists had been examining as a precursor to more involved pyrimidines, part of the D.N.A. structure. Because of its more complicated formula, cyanoacetylene was an even more important molecule to find than hydrogen cyanide (HCN), which had been detected in space a few months previously.

Many of the complex molecules found in space have names that are unfamiliar to the layman—but the relative complexity of their chemical formulae should be impressive enough to anyone. Among these advanced compounds built up by natural processes in deep space are methyl alcohol (CH_3OH), formamide (H_3NCO), methylacetylene (CH_3C_2H), methyl cyanide (CH_3CN), acetaldehyde (CH_3CHO), formaldimine (CH_2NH), ethyl alcohol (C_2H_5OH) and dimethyl ether, formula $(CH_3)_2O$. Two of the molecules found in space, formic acid (CH_2O_2) and methylamine (CH_3NH_2), can react together to give the amino acid glycine.

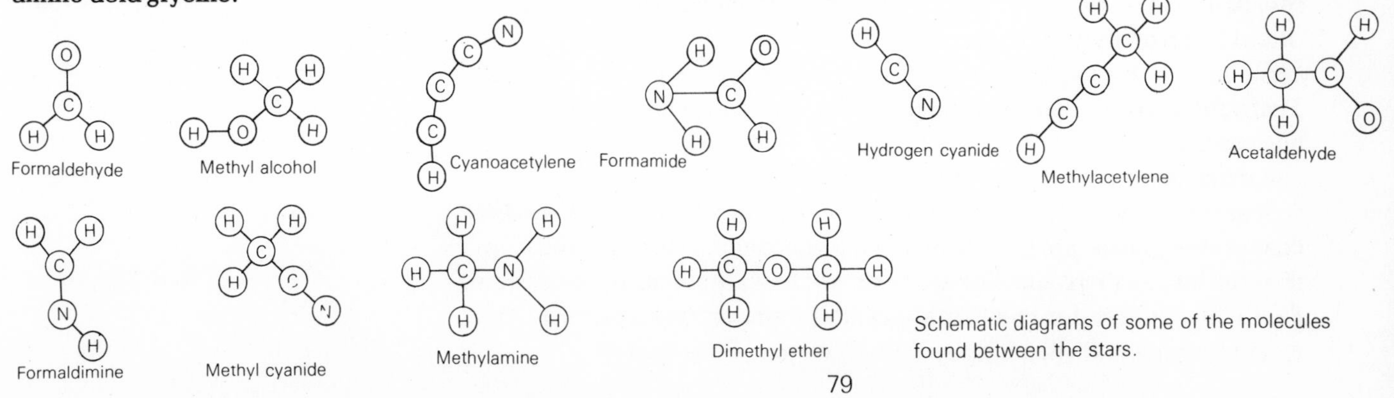

Schematic diagrams of some of the molecules found between the stars.

Interestingly, astronomers have detected sulphur in space, first in the molecule carbon monosulphide (CS), and in 1972 in hydrogen sulphide, H_2S—the gas that gives rotten eggs their bad smell.

Not all the identifications have been made so easily. For some time astronomers puzzled over the emission from a substance they labelled X-ogen in admission of its unknown nature. Only in 1974, four years after its discovery, did it become reasonably well established that X-ogen is an electrically charged combination of hydrogen, carbon and oxygen.

In a paper in 1972 entitled 'Interstellar Organic Chemistry', in the journal *Nature*, Carl Sagan discussed the significance of these results. Pointing out that the molecules discovered between the stars are 'strikingly similar' to those observed in comets and in experiments on pre-biological organic chemistry, he noted that the conditions inside a contracting gas cloud round a young star are similar to those on the primitive Earth.

And he asked, can interstellar organic compounds make a significant contribution to the origin of life on a planet formed from such a cloud?

The answer, he believed, was no—because the yield of such compounds in primitive atmospheres is far greater than all the possible 'seeding' effects from space, including meteorites. The tempestuous geological history of the primitive Earth would in any case have destroyed such early deliveries of organic material.

Once again, therefore, even the updated ideas of panspermia become scientific non-starters.

Instead, Sagan preferred to regard the build-up of molecules in space as analogous to the processes that led to life on Earth—and perhaps on vast numbers of planets throughout the Universe.

He also considered the possibility of an interstellar organism living among the clouds of space. For simplicity, he envisaged a creature the size of the smallest known organisms on Earth.

He found that there would be so few encounters with the thinly spread material in a cloud that even a tiny creature would take around 200 million years to mop up enough extra material to reproduce itself.

This means that there can have been only about 60 generations of such an organism in the history of our Galaxy—far too short a time, says Sagan, for it to have evolved through natural selection to a stage capable of functioning in so inclement an environment.

In his paper, Sagan took the view that the observed compounds in space could be due to the breakdown of still larger organic molecules—hinting that substances such as the amino acids found in meteorites would also be proved to exist in the interstellar clouds.

A searcher for molecules in space, American radio astronomer Patrick Thaddeus, is equally optimistic about the exciting prospects for interstellar chemistry. He says, 'I would not be in the least surprised if many dozens—even hundreds—of molecules were eventually found in interstellar space.'

There remains one important point to consider: do we have any right to assume that life-forms elsewhere in space would be based on the same chemicals as ourselves?

The ubiquity of carbon in the interstellar molecules so far discovered, particularly the most complex ones, is encouraging support for the idea that life elsewhere should be carbon-based, as it is on Earth. The reason for this is that carbon is an atom with a particular affinity for joining with others.

But there are other possibilities, one of the most famous of which is a theoretical life-form that would be based on silicon instead of carbon. Silicon is a type of atom that easily makes complex molecules; the rocks of the Earth are based on silicon. The problem with silicon is that of finding a suitable solvent. Water will not work, because in its presence silicon rapidly turns into silicon dioxide, which we know as the mineral quartz. A silicon-based life-form may be possible, but biochemists are not very confident that it would occur naturally.

Another possibility is life in which carbon combines with chlorine instead of hydrogen. Although the resulting chemicals are rather exotic, biochemists are satisfied that such a system is theoretically possible.

Perhaps the most closely examined alternative biochemical system is one which uses liquid ammonia instead of water as a solvent. This liquid-ammonia type of life, which again is carbon-based, is a candidate for existence on colder planets. Peter Molton argued strongly for liquid-ammonia life in a paper published in 1974 in the *Journal of the British Interplanetary Society*, in which he showed that the major molecules of life could be successfully constructed using ammonia instead of water. 'The giant solar system planets may support life of this kind,' he says.

There may be many other possible combinations that could give rise to life, but biochemists remain so ignorant of the chemical complexities involved that they are unwilling to bank on any such systems.

But, surveying the possibilities, most biochemists return to carbon-based life with a water solvent as the most probable combination: both because we know it works well, and because among all the alternatives these substances are the most abundant.

As we have seen, the clouds that produce stars are also those that produce the first molecules of terrestrial-type life. So it is not just chauvinism to assume that the Earth's inhabitants are typical of the biochemical forms that the Universe has to offer. It is also practical science.

ORIGIN OF THE SOLAR SYSTEM

Five thousand million years ago the first dawn had not yet broken on Earth. No Sun scattered its spark into space, and there was no Earth to spin in its warming rays.

Our region of space resembled nothing more than a thin, freezing fog, lit only by the glow of distant star-fires. But out of the swirling gases of that icy fog sprang a star with a brood of planets. The young Sun's incandescence beamed upon this growing family and moulded them into shape.

In the sky of the embryo Sun, the distant star-fires showed that the same events were repeated many times over, as they had been throughout the preceding 5,000-million-year history of the Galaxy.

Yet on one of those siblings arose a civilization of inquisitive creatures who looked into the sky and tried to reconstruct the events of their distant past. Legends of the birth of the Earth and sky are as old as thought itself. It is from this urge to invent a creation that the first religions spring. The legends traditionally emphasized the Earth's supposed special place in the order of things, an attitude that did not materially alter until the seventeenth century, when astronomers demonstrated that the Earth is not the sky's central body.

The attitude was finally undermined earlier this century by the realization that we are not even centrally placed in our Galaxy, and that the Galaxy is an unimportant member of the Universe.

One of the first truly scientific ideas of the Solar System's origin was put forward over two centuries ago by the German philosopher Immanuel Kant. In his view, spinning disks of gas formed from a thin medium that initially filled all of space. At the centre of the disk, like the hub of a grindstone, formed the Sun. The planets were assembled from the material that surrounded it. Kant also believed that whole star systems would follow the same pattern, and he wrote of 'island universes', which we would now call galaxies, that formed lens-shaped patches in the sky.

Later in the eighteenth century, the French mathematician the Marquis de Laplace proposed a more detailed version of the same idea. Laplace assumed that a large cloud of gas was drawn together to make the Sun. As it did so, it left behind a series of rings as the centrifugal force of its rotation overcame the Sun's gravitational attraction. From each of these rings formed a planet.

In essence, such ideas provide a remarkable preview of contemporary views about the origin of the Solar System. But for much of this century these so-called nebular theories were out of favour, and astronomers looked for different answers to the puzzle of the planets' existence.

One idea examined in some detail was that the Sun may have drawn round it a cloak of material as it passed through one of the many clouds of dust and gas in the Galaxy. Astronomers know that such clouds are common; an encounter with one would be a likely event in the history of a star.

The captured material would smooth out into a spinning disk, so the theory went, and from this disk would form the planets. Yet doubters have claimed that the Sun could not easily capture a suitable mass of material. More likely, they think, it would fall straight into the Sun or swiftly disperse again.

Perhaps the most famous theory of all, because of its splendidly catastrophic nature, was the collision hypothesis that envisaged a near-miss between the Sun and another star. Sir James Jeans was the great popularizer of this idea in the 1930s. Variations on the same theme were proposed by several astronomers over many years, yet they all rely on an encounter between the Sun and another star—which, because of the vast distances in the Galaxy, would be such an unlikely event that very few planetary systems could be formed in this way.

As the two stars came within hailing distance, Jeans envisaged, giant tides were raised on the surface of the Sun and a cigar-shaped filament of glowing gas was drawn out. This swung into orbit round the Sun, and the material in it condensed into planets—the largest ones at the centre, and the smallest at the ends, as observed. But modern astronomers now calculate that droplets torn from the Sun would not fall into orbits like those of the planets, nor would such material form objects of the right chemical composition.

Most damaging of all, theorists have realized that such a filament of hot gas would never condense into small, solid bodies. Because of its heat it would simply disperse into space. In its stated form, the collision hypothesis could not produce a planetary system.

A variety of other suggestions to explain the Solar System has since been made, including the possibility that the material of the planets is left over from a disrupted companion star of the Sun.

One staggering new development came in 1972, when Wilbur Brown of the University of Wyoming proposed that new star-and-planet systems are formed directly from the ejected shells of supernovae. Brown believes that such an ejected shell of supernova material will fragment into lumps which subsequently evolve into several new solar systems. And he makes calculations showing that this theory will explain many of the observed properties of the Solar System.

'The model gives a remarkably good description of our solar system, in both its broad regularities and its specific anomalies,' he claims.

But, so far, his idea has received little support from other astronomers.

The following year, two theoreticians of York University, C. Aust and M.M. Woolfson, showed that the collision hypothesis is not yet totally out of the running. They affirmed that the Sun could have captured enough material for a planetary system from the cool outer layers of a forming star. The gas could have been pulled out into a tongue, fragmenting into condensations which were captured round the Sun as precursors of the planets.

'Such an event might be likely in the early days of a star cluster,' says mathematician Iwan Williams, who has made a major survey of theories of Solar System origin.

Yet Williams, like many other theoreticians, is now inclined to the belief that this somewhat chancy process was not how the Solar System formed.

Instead, the majority of present-day astronomers prefer the view that the origin of planets is intimately bound up with another major process in astronomy — the birth of stars.

Until very recently, scientists have been little better off in their understanding of star birth than they have been in unravelling the origins of life. They have the major and insuperable disadvantage that they are unable to test their theories by experiment.

But, thanks to the theoretical models that can be worked out using high-speed computers, the general processes leading to star formation now seem clear, although the confirmatory observations remain somewhat sketchy.

A star begins, most astronomers agree, as a widespread cloud of dust and gas in the spiral arms of a galaxy. The material of which it is made is the primordial mixture of hydrogen and helium gas, seasoned with heavier elements disgorged from aged stars — a process we reviewed in Chapter One.

Yet such a cloud on its own is not an inevitable candidate for the new-star stakes. We see many such clouds in the Galaxy which are apparently quite happy the way they are. They show no signs of forming stars. To form into a star, something must nudge the cloud so that it starts to collapse in upon itself. Then its own gravity will pull it together.

Astronomers have suggested several ways in which collapse may be started off. They think that big swirls could be set up in the cloud by magnetic fields in the Galaxy, or by heating from cosmic rays or X rays.

According to one now-popular theory, there is a constant wave of gravitational force that sweeps round the Galaxy, which is responsible for setting up the spiral arms. As this wave sweeps through gas clouds,

The remains of an exploded star drift into space, eventually to be swept up into new stars and, perhaps, planets. (*Hale Observatories*)

A dark lane of dust rims a spiral galaxy seen edge-on to us in space. (*Hale Observatories*)

it causes material in them to bunch up. Where the clouds become dense enough, they will start to collapse into smaller, opaque spheres, inside which a young star will eventually form.

'This picture of star formation seems to have very strong observational backing,' says the University of Arizona's Bart Bok, a lifelong researcher into star formation. 'The various stages are as plainly visible to the eye as plants in successive stages of development in a garden.'

Yet however star formation starts in the galactic garden, the theorists calculate that a collapsing cloud must initially have a mass at least a thousand times that of the Sun—or else it would not have enough gravitational pull to get going. In other words, individual stars are never formed. They must come into being as a complete star field. 'Star formation proceeds via the fragmentation of a massive collapsing cloud into smaller subcondensations,' says Yale University Observatory's Richard Larson, who has performed computer simulations of the birth of stars.

Astronomers point to the Coal Sack nebula in the southern hemisphere as an example of just such a process in operation. Smaller, darker blobs can be seen within the Coal Sack's sooty spread. 'The Coal Sack nebula looks like the place where a star cluster is about to be born,' Bart Bok enthuses.

Before it starts to collapse, such a cloud has a temperature of between about 20° and 60° above absolute zero—the very coldest temperature possible, which is 273°C below the freezing-point of water. Even at this temperature the cloud is quite warm in comparison with empty space, because it is being heated by cosmic rays and X rays. But, as parts of the cloud get denser, these heating agents are shielded

Arcturus
Epsilon Boötis
North celestial pole
Centre of our Galaxy
Vega
Barnard's Star
Altair
Andromeda galaxy
Pleiades cluster

This panoramic view of the entire sky seen from Earth shows the dark lanes of dust running through our own Galaxy. Also marked are some of the objects mentioned in this book. Because the picture is centred on the line of the Milky Way it differs in appearance from other maps of the sky. (*Lund Observatory*)

out and the cloud's temperature actually falls—this time to about 10° above absolute zero.

This is not the popular picture of a forming star, but the work of radio astronomers supports the theorists. Those complex organic molecules observed in the interstellar clouds tell that the temperatures and densities are exactly as predicted. Indeed, even until the moment when the star first switches on, the temperature in the cloud round it rises to no more than a few hundred degrees.

At several places inside the low-temperature cloud the density becomes high enough for masses of dust and gas the size of the Sun to collapse. The cloud then fragments into the individual condensations that directly spawn stars. Throughout the Galaxy, astronomers can see small, dense clouds that are collapsing on themselves at a rate of about one kilometre every second; just the rate that is expected for a forming star. A single star on its way to becoming a recognizable, glowing object is thought to look like the tiny dark globules discovered by Bart Bok, and which are named after him. Such Bok globules seem to be about the same mass as our Sun, and to occupy an area not many times larger than our Solar System.

Astronomers expect that such a globule will be slowly rotating. There is no way to get rid of this rotation, so it remains trapped in the collapsing globule.

As the globule compresses itself, the rotation is compressed too. If the globule is big enough, it may end up spinning so quickly that it bursts like a flywheel, flinging matter out into space. Among this matter may be other condensations that later turn into individual stars. About half the stars in the sky are seen to be double or multiple, and this is one way of explaining their frequency. But another point of view, as we shall see below, suggests that multiple stars can be formed from the left-overs after a single star is born.

What is the spark that kindles a star in the still-cool globule? The vital spot is at the globule's centre, where the density of the collapsing matter goes up most rapidly. Most of the mass is left behind, forming a dusty envelope that collapses much more slowly—a reluctant follower of the downrushing core.

As the overall density of the globule increases, and the dust and gas particles rub together, the globule can no longer radiate away all the frictional heat. At last it begins to warm up, and disperse the icy fog. The object has reached the stage that astronomers term a *protostar*.

Astronomers can detect the heat of such forming stars at infra-red wavelengths. They believe that there are a number of embryo stars in part of the Orion nebula, a famous mass of gas which glows to the naked eye as a result of the luminous stars that have already formed within it.

As the material of the surrounding globule cascades down upon the surface of the protostar, its temperature quickly rises to many thousands of degrees, and it glows more brightly than our own Sun does today. Although a globule may take a million years to first form, once it has begun to contract it may be but a few more years before the protostar is first seen to glow. The energy of its gravity then powers it for around 10 million years.

This energy is eventually replaced by the steadier glow of the nuclear fires that are tindered by the extreme temperature and pressure at the young star's heart.

Evidence for a collapsing cloud's sudden rush to stardom was provided in 1954 when astronomers photographed a glowing clump of gas in Orion that had not been visible on pictures of the region taken a

1 The dark Coal Sack nebula, near the Southern Cross, is a possible breeding ground for stars in our Galaxy.

2 Dark lanes and large globules among the stars of the Milky Way. (*Hale Observatories*)

Smaller globules show up dark against a brightly glowing nebula. (*Hale Observatories*).

1
2

few years before. They interpret this as the observation of a new star that has just been switched on by its gravitational energy.

To confirm their ideas, they eagerly await the appearance of similar protostars at the infra-red sites that they believe are collapsing globules.

Another class of objects, called T Tauri stars after the first one of its type to be discovered, seem to be young stars still embedded in the swirling, cloudy envelope that gave birth to them. Astronomers believe that these objects are the visible link between dark globules and fully formed stars.

And they are important because they seem to be shedding material as the Sun is believed to have done in its early days—a remarkable confirmation of the visionary work of the Marquis de Laplace. As Laplace foresaw, a shrinking star is likely to be rotating so swiftly that the gas at its surface reaches escape velocity. A stream of proto-sun material spirals away into the surrounding cloud, and flattens out into a spinning disk round the youthful star. Here is the potter's wheel on which the planets will be thrown.

'The theory for the formation of a protostar from a dust cloud seems to lead almost naturally to the formation of a planetary system,' says Bart Bok. It is this realization that has excited the scientists who search to explain our origins—and boosts their optimism that we are not unique.

An interesting twist to this accepted tale was added in mid 1974 by the Soviet astrophysicist Edward Drobyshevsky. Writing in the magazine *Nature*, he contended that the outer regions of a forming protostar would be so unstable that almost all of its material would be thrown into a ring. And it is from that ring that the major component of the system—the future star—would be formed.

By this theory, Jupiter becomes the core of the dismembered protostar, and the Sun is formed from the majority of the material thrown off.

The small planets between the Sun and Jupiter were formed when matter streamed between the two components, says Drobyshevsky, and the outer planets were formed from the remaining gas that was spun off from the ring. In any case, either a double star or a planetary system is the natural by-product of this process. But the idea is so new that it has not yet received full consideration from other astronomers.

The Chinese-born astrophysicist Su-Shu Huang of Northwestern University, Illinois, believes that in such events as the switching-on of stars in Orion, we are observing the first stages in the birth of planetary systems. As the newly glowing stars eject matter into a ring round them, the remains of the surrounding gas envelope collapses into a disk as the result of its high-speed rotation.

The next stage remains the most disputed among theorists. Almost certainly, they now believe, the material inside the star produces a magnetic field that seeps into space. And a wind of atomic particles, blown off the incandescent surface of the energetic young star, curves outwards along these magnetic lines of force.

The particles that gush outwards from the swiftly spinning star carry its rotation into the cloud. The effect is to brake the spin of the star, and to wind up the spin of the disk of dust and gas. This helps explain one of the observed oddities of the Solar System: most of its mass is concentrated in the Sun, while most of the rotation is spread among the planets.

Without some form of brake on its spin, the Sun would be expected

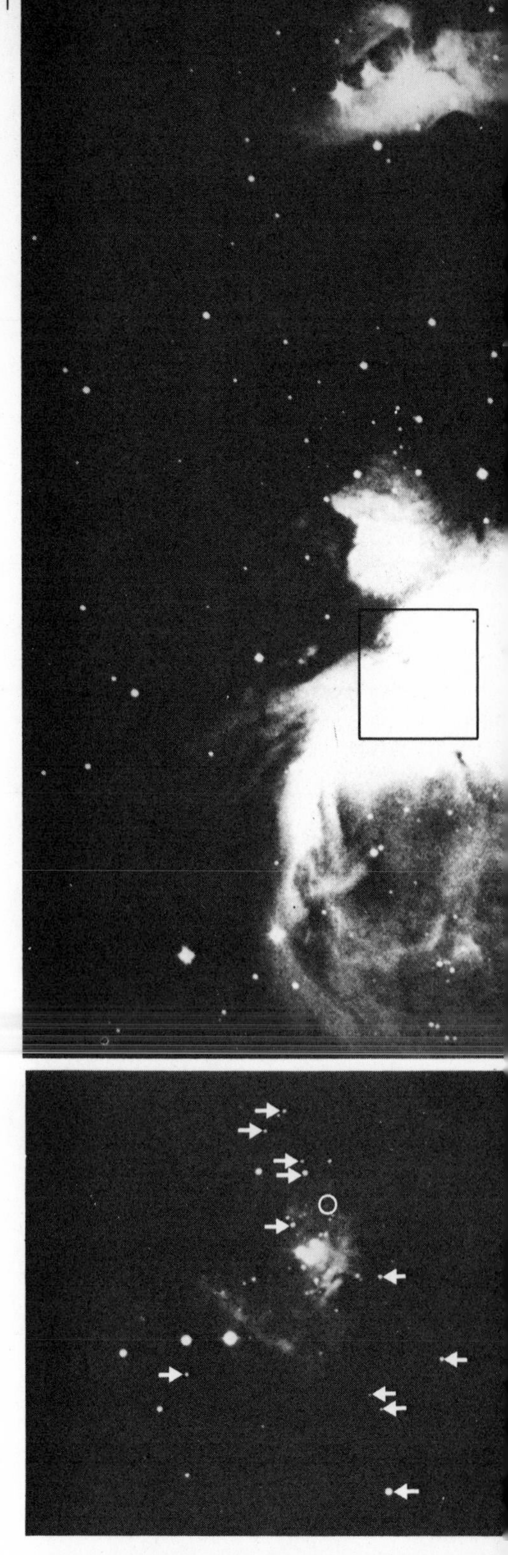

1 The Orion nebula is the glowing front to a cloud of dark and dense gas that lurks behind. The view of the complete nebula is shown with a square marked on its centre. The smaller picture, taken in the near infra-red, shows this central portion of the nebula in detail, with young stars (*arrowed*) and the site of a protostar (*circle*). (*Royal Observatory Edinburgh, and Lick Observatory*)

2 The Pleiades are a group of stars visible to the naked eye as a fuzzy patch in the constellation Taurus. They are still surrounded by remnants of the dust and gas from which they were born. The Pleiades are thought to have formed within the last 200 million years. (*Lick Observatory*)

3 In this cloud of gas, called the Lagoon nebula, stars on the left have recently formed, while the process of star birth seems still to continue in the nebula to the right. (*Hale Observatories*)

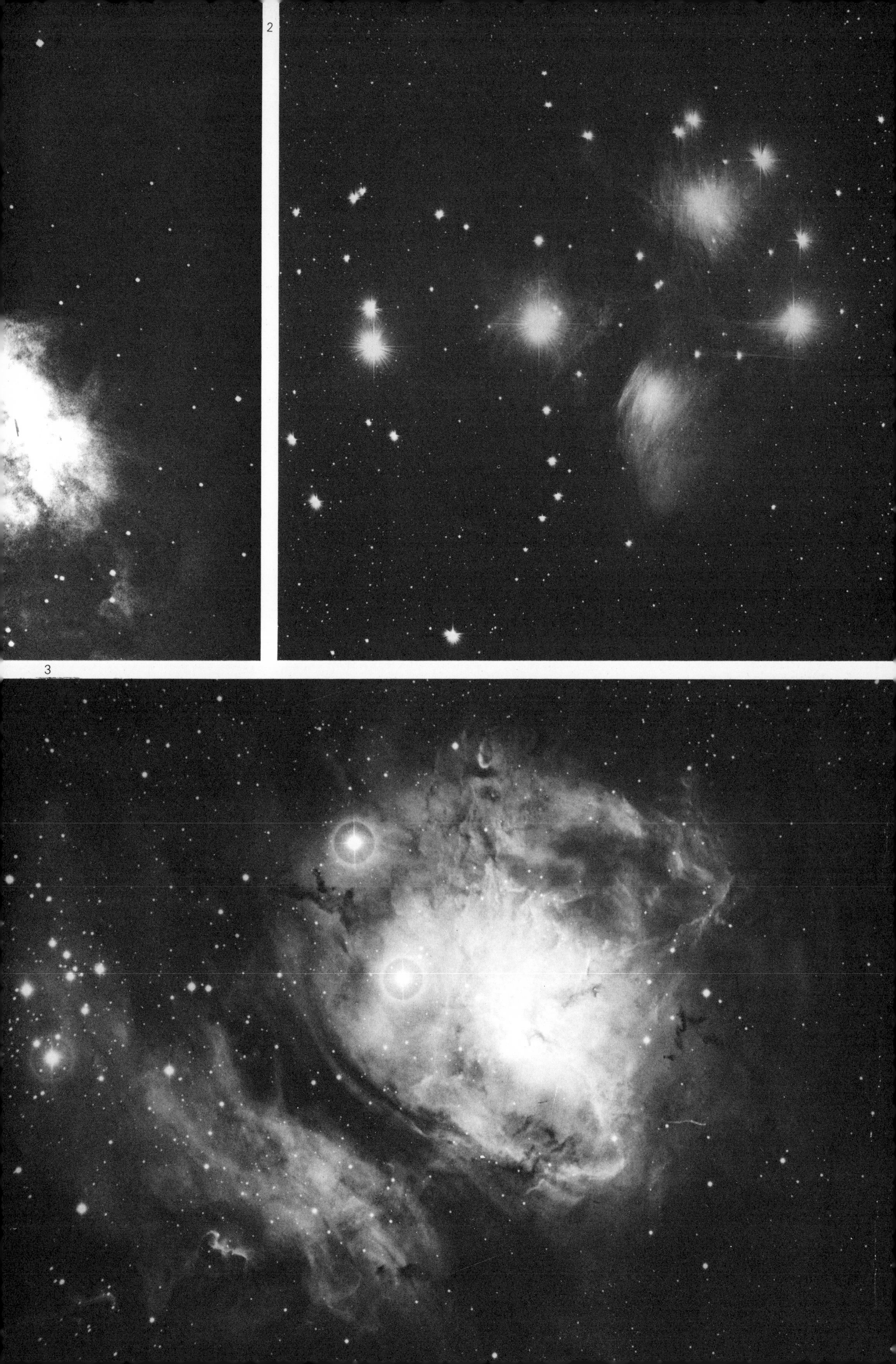
2
3

to turn once every few hours, rather than the observed once a month. Equally, the majority of the other single stars in the sky seem to spin far slower than would be expected if they had formed from a collapsing cloud of gas. Astronomers believe that these stars have also been disk-braked, which Huang considers is another pointer to the frequency of planetary systems in our Galaxy.

Huang is also struck by the fact that the slow-spinning stars, like the Sun, are those which live the longest. 'Planetary systems appear just round those stars that can support life,' he says.

In the case of our own Sun, calculations suggest that it would have begun to shed its spare-tyre material when it had condensed to about the size of the present orbit of Mercury. The transfer of rotation from the Sun would have spread this material into a disk extending throughout the present-day Solar System.

This disk of dust and gas is termed the solar nebula. Unlike the filament of gas that is drawn from the Sun in the collision hypothesis, this nebula would not be incandescently hot. The Sun's outer layers would still be relatively cool, so that the material spinning off into the disk would be no hotter than 1,000° and in many places much less.

'The primitive solar nebula was obviously a highly complex system,' says astrophysicist Alastair Cameron of New York's Yeshiva University, who with a colleague, M.R. Pine, has attempted to calculate some of the effects that lead to planetary formation in a spinning disk.

Theorists still argue about the exact events in planetary build-up. But the consensus of views is as follows.

Inside the nebula, specks of dust collided and stuck together to make larger lumps. Each dust speck, like the grains in interstellar clouds, was a smoke-sized particle formed of many millions of the atoms that make rock and metal.

The process by which the dust motes stuck together is called *cold welding*, which scientists can duplicate in the laboratory. The grains aggregated into larger lumps, perhaps a foot or two across, which formed a carpet through the plane of the Solar System. Each lump, on its own orbit round the Sun, would grow by collisions with other lumps until it was big enough for its own gravity to pull other lumps towards it. By this random process, meteorite-sized fragments and then asteroid-sized planetesimals would form. And the time-scale would be very short: these emerging planets would grow in a matter of a few thousand years, the theorists calculate.

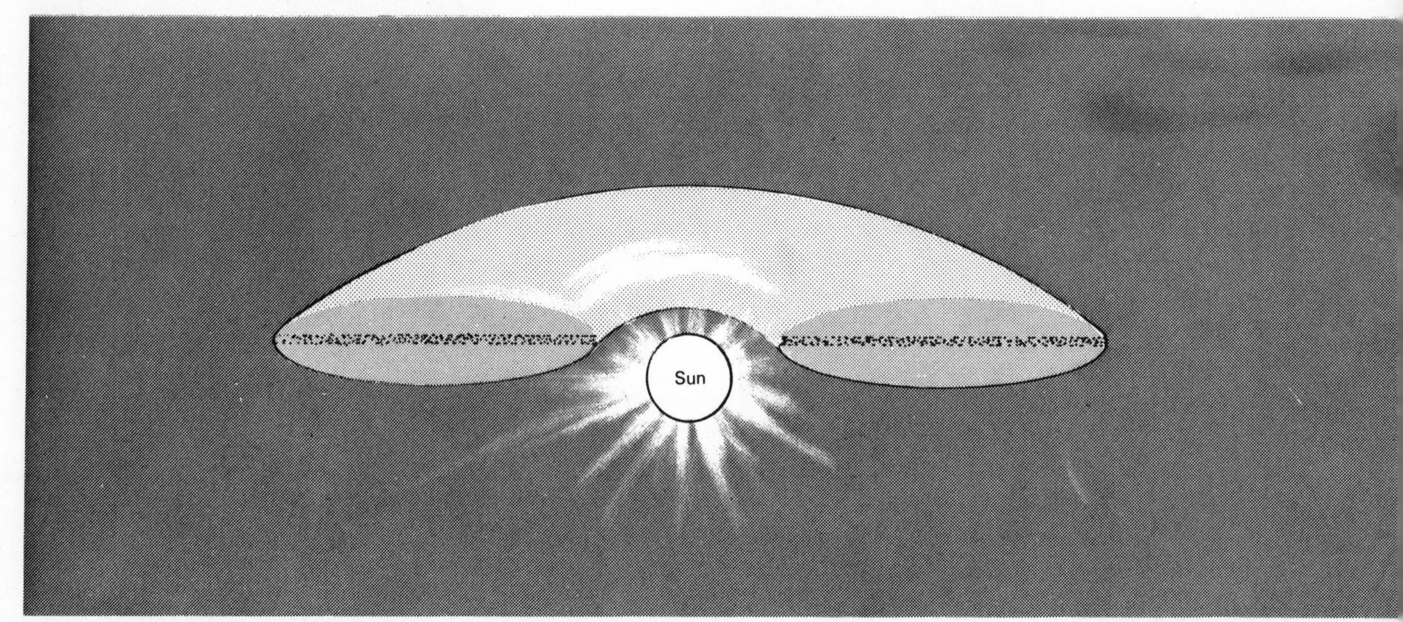

How the Solar System started to assemble: a carpet of dust grains, surrounded by a doughnut of gas

In their earliest stages, then, the solid cores of the planets were subjected to a continual bombardment as the major masses of the Solar System were swept up. In some cases, large-sized lumps were captured into orbits round the young planets, where they remain as natural satellites.

The largest planets of all—notably Jupiter and Saturn—were able to draw in their own mini-nebula of dust and gas, which developed by similar processes into a complex system of satellites, like a smaller version of the Solar System.

There were other events in the nebula that helped shape the final form of the planets. One of these was the effect of a flow of atomic particles called the solar wind, which blew much of the nebula away. The planets closest to the Sun were built up of the heaviest material, which was less easily pushed away from the Sun. These planets have a core of heavy metal sheathed by rock. Most of the gas in the solar nebula was quickly pushed out into the depths of the Solar System, where it could be captured by the giant planets forming far from the Sun. But at the very edge of the Solar System, the gas would disperse before it could form into planets.

As many as 99 atoms out of every 100 that spun off from our shrinking star were lost into space for ever. Mostly these were atoms of the lightest gases, hydrogen and helium.

The first atmosphere grabbed by the gravity of the Earth was made up from the traces of gas that had not been spun away from the Sun. Here was ammonia, methane and water vapour, plus those more complex organic compounds that radio astronomers detect in the clouds round forming stars—a brew that, as we have seen, the chemists believe could have formed the first complex molecules of life.

'The condensation of a star, the accumulation of the dust and molecules into planets and atmospheres, and even the subsequent evolution of life may all be part of an astronomical evolutionary cycle of very long time scale,' says radio astronomer David Buhl, a discoverer of many complex compounds in space.

In time, the heat of the warming Sun would have evaporated this primitive atmosphere of the Earth—a type of atmosphere which the other inner planets may also have possessed. But, simultaneously, the flaring volcanoes of our young planet were exhaling the gases of our secondary atmosphere—including the water vapour that condensed into rain and filled the seas. Though the planets were cold when they formed, the natural radioactivity inside them would soon have heated their interiors to melting temperatures.

So here, supplied by the work of astronomers and geologists, are all the conditions that the biochemists believe they need in order to explain the origin of life.

Looking back on this period, the Swedish astrophysicist Hannes Alfvén, a keen investigator of the Solar System's early history, has said, 'The situation on the Earth was probably rather dull, because it had been raining solid particles continuously for a few hundred million years. It was not a heavy rain, but it gave a precipitation of an inch or two a year.'

In diminished form, that same rain still continues, as the Earth sweeps up a few hundred tons of micrometeorites every day—remnants of the solar nebula that have been on the loose for 4,500 million years.

The planets farther from the Sun have changed very little since their origin. Jupiter, Saturn, Uranus and Neptune are in the icier outer regions of the Solar System where the Sun's heat had less effect. Their atmospheric gases have not evaporated, and they provide a literally frozen picture of how the Earth might have looked soon after its birth.

And farther out still, the comets probably represent the icy conglomerations from which the outer planets first formed. The organic component of a comet's gases, as well as the complex compounds found in meteorites, provide us with well-preserved samples of the solar neighbourhood at the time of the Earth's genesis.

To check the outcome of these processes, an American computer engineer, Stephen Dole, has run a series of simulations of planetary build-up from the solar nebula. His results are truly revelatory.*

Dole starts with a Sun-sized star surrounded by a disk of gas and dust similar to that which gave birth to the planets. Inside this cloud he imagines that the dust-grains on their orbits round the Sun grow by random collisions, as the modern theories envisage, until a suitable nucleus for a planet is built up. The computer then works out what happens next. Where two such planetary nuclei come close enough to be gravitationally attracted, or where their paths cross, they coalesce into one larger body.

As these growing nuclei orbit within the cloud, they sweep out dust-free lanes. The largest objects so formed can also draw in gas. Such nuclei are formed throughout the cloud, on orbits of varying size and shape. The process continues until all the dust is swept up and the left-over gas is driven out of the system by the solar wind.

A print-out records the final result. 'The computer runs produced planetary systems with all the general characteristics of the solar system,' says Dole in summary of his results. Despite varying the conditions slightly between each run, the planet-building process resulted in small rocky planets closest to the Sun, large gas planets in the middle of the simulated Solar System, and small planets again in the outer orbits.

There were between 7 and 14 planets formed in each system. The total mass of the planets was similar to that of our own Solar System, and the largest planet in each case resembled Jupiter in size.

All this tends to make our Solar System look remarkably average—which is a boost to those astronomers who maintain that we are just an ordinary example of a common process within the Galaxy. Even planetary experts are unable to pick out our own system immediately when a series of Dole's computer runs are displayed.

One finding, which confirms a long-held suspicion among students of the Solar System, is that under certain conditions a small star can form in orbit round the Sun. In Dole's simulations this occurred when the density of the nebula was increased, thereby allowing more material to pack together. Had this happened round the Sun, Jupiter would have become a dwarf star, the planetary scientists maintain. And in a big enough nebula, perhaps even a three-star system could have ensued.

'The aggregation theory of the formation of planetary systems can account for many of the intriguing regularities and irregularities of the solar system, and is compatible also with the observed properties of many of the known, widely separated binary star systems,' Stephen Dole concludes.

Can planets survive in double-star systems? This question has been examined by Shiv Kumar of Leander McCormick Observatory in Virginia. He finds that, if the mass of Jupiter were greater or if its orbit were less rounded, or both, then the orbits of the smaller planets in the

* *Icarus*, 13, 494, 1970.

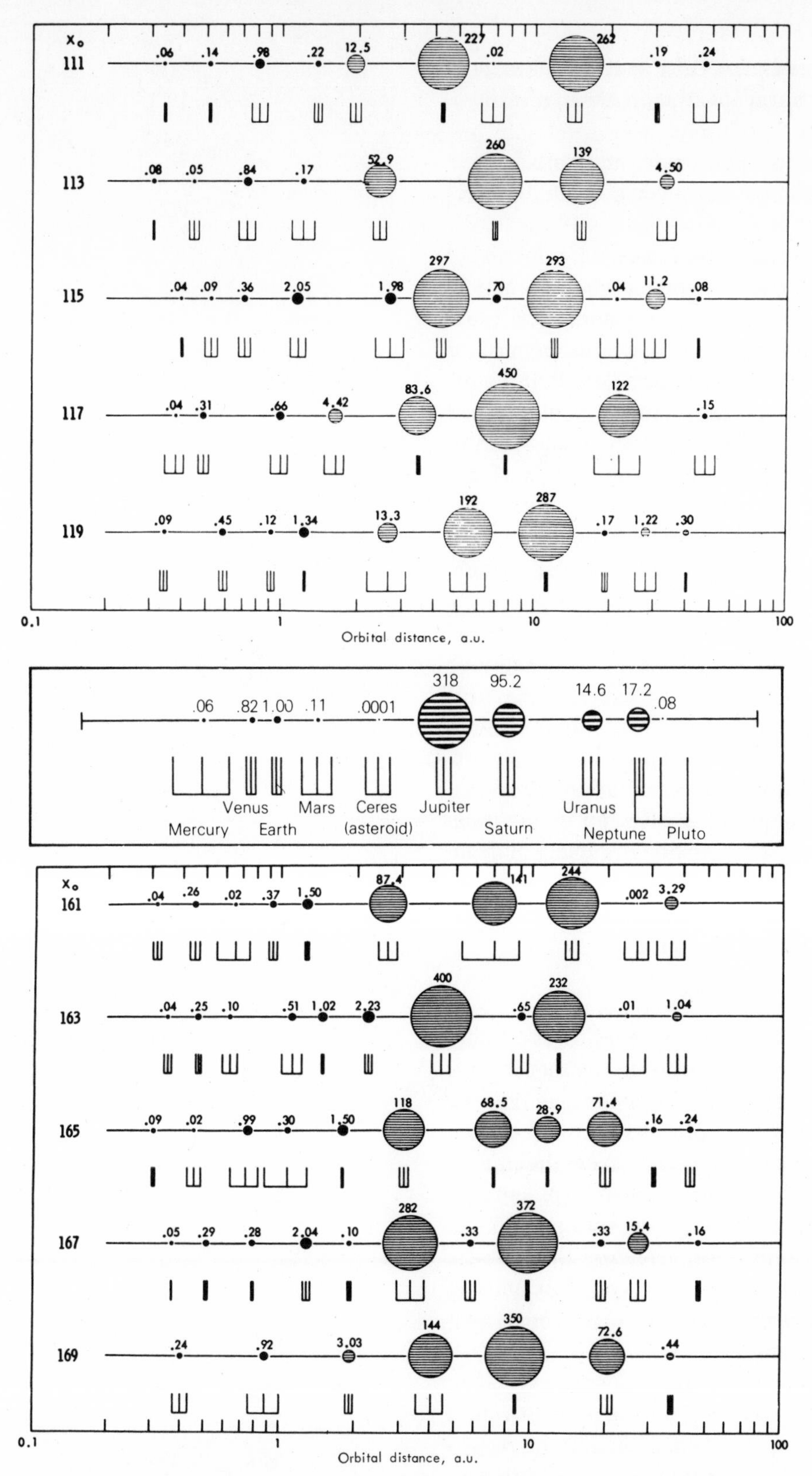

Solar System simulations by computer. Figures at the left indicate conditions in the forming nebula. Figures at the foot are distances from the Sun in terms of the Earth's distance. Figures above the planets are masses in terms of the Earth's mass. Brackets below the planets indicate the eccentricity (degree of ellipticity) of each orbit. Compare our actual Solar System, drawn to the same scale and style. (*Stephen Dole*)

Solar System would become unstable. These planets, including the Earth, would collide with either of the stars, or they would be thrown out of the system altogether. Kumar therefore concludes that many of the observed double-star systems, which contain high-mass stars with eccentric orbits, 'cannot possess planetary systems such as our own'.

However, Kumar's work revives a supposition made by some astronomers that there may be planets floating free in space, attached to no star. The noted astronomer Harlow Shapley, in addition, had a favourite theory that Lilliput stars existed, which were sufficiently cool for life to form on their surface or in their atmospheres. Biologists, though, are not very impressed by the prospects for life on a sunless planet or a dark star. There are problems: lack of a source of energy to power the life processes, and the general low temperature of the environment.

While life-forms in such situations are not totally impossible, their likelihood seems low. And we may in any case never be able to detect such unlit habitats.

One fact about our own Solar System that has long intrigued astronomers is the harmonious spacing between planetary orbits, almost as if they had been placed according to a mathematical plan. This characteristic is partly present in Dole's computer-generated solar systems. Astrophysicist Thomas Gold has pointed out that the orbits of the computer-planets are less circular than the orbits in our own Solar System, and expresses the opinion that they would tend to become more rounded if the computer simulations were run on for a much longer time.

In 1969 the American theorist J.G. Hills examined, mathematically, a number of imaginary solar systems with sizes and orbits of planets picked at random. He found that the effect of each planet's gravity would cause the others to move into orbits where they least affected each other. The result in each case was that the orbits became spaced out in similar fashion to what we find in the Solar System today. He further concluded that the major planets in our own Solar System would have taken up their positions within a million years of their formation. The fact that the Solar System is so gravitationally stable today, with the orbits of planets being almost unaffected by the pulls and pushes of their fellow worlds, is itself a strong argument that no catastrophic events have happened in recent Solar System history.

There is, in fact, every reason to believe that the Solar System has survived almost unchanged since soon after its birth.

If solar systems form so readily, what is the evidence for their existence elsewhere? Unfortunately, the obvious way to find out—simply by looking—doesn't work. From another star, even the largest and brightest planets in our own Solar System would appear as nothing more than the faintest dots, at the very limit of detection in even the largest telescopes. And, in practice, the dazzling light from the parent star would swamp the much fainter pinpricks of any nearby planets.

Things may be improved when large telescopes are set up in space, above the blurring effect of the Earth's atmosphere. But that is technology for the future, and we can find other ways of examining stars from the ground, which is where the majority of astronomy will continue to be done well into the next century.

One promising method is to measure light from a star over long periods of time to see if there are periodic partial eclipses that would be caused by a planet moving in front of the star. Although the reduction in the star's light would be small, we could detect it by sensitive instruments on the ground, and several schemes have been proposed for doing so.

They are hampered by the fact that only a small proportion of solar systems will be suitably aligned for us to see such disk traverses. Yet the late Professor Frank Rosenblatt of Cornell University believed that a system could be devised at reasonable cost that would be likely to detect at least one planetary system per year. One of his detailed proposals was a group of three telescopes searching a total of nine thousand stars each night for the type of star-darkening that would be caused by a planet of Jupiter's size. When such a system was detected, it could be examined in more detail for the presence of smaller companions. But the job would be exhausting: a planet such as Jupiter orbits the Sun every 12 years, and it could be seen crossing the face of the Sun for only one day during that time.

An American astrophysicist, Gregory Matloff, has pointed out that the nature of a planet could be given away by its colour. Planets like the Earth are bluer than their primary star, while objects like Mars, Venus or Jupiter appear redder. Matloff believes that, in principle, observations from a large space telescope would allow us to distinguish between planet types in another solar system.

Canadian astronomer Edward Argyle added to the list of possible planet-spotting techniques in 1974. He pointed out that the light from short flare-ups of faint stars would be reflected to Earth from any planets orbiting that star. This would produce a secondary peak of light emission. But it is an event that would be observed only infrequently.

Observing at optical wavelengths may not be the best way to detect planets of distant stars. The difference in luminosity between stars and planets is reduced at infra-red wavelengths, and British communications engineer Tony Lawton is one who has proposed an infrared telescope system for monitoring nearby stars with suspected dark companions. Such a giant telescope, which might be as much as 300 feet across, would be best sited on the far side of the Moon.

There remains one other method for detecting the planetary companions round stars. It is the most indirect method of all, for it involves watching the deviations of stars from their smooth traffic pattern round the Galaxy. But it has produced tantalizing results.

As the Galaxy spins, the stars within it whirl through space. The Sun takes about 250 million years to go once round the Galaxy. This means that the Solar System has completed less than 20 circuits since its birth. The Sun is lapping at a speed of about 160 miles per second, but the speeds of other stars differ, depending on where they are in the Galaxy. Objects closer to the centre orbit the quickest; those farthest out go most slowly.

The situation as seen from the Solar System is like being in the middle lane of a wide highway. Vehicles in the slow lane seem to drop behind; in the fast lane, cars move ahead. Similarly, the smooth orbits of stars should take them ahead of or behind the Sun. By photographing the stars over long periods of time, astronomers can measure their differences of speed. Over many thousands of years, these speed differences mean that the shapes of constellations change radically. But, in a human lifetime, the differences are too small to detect with the eye alone.

Instead, astronomers use telescopes to take large-scale pictures on which they measure star positions with highly accurate machines. This technique is called *astrometry*, and is a slow, laborious procedure—not nearly as glamorous as the exciting results it can produce.

The prime exponent of planet-finding by astrometry has been the genial Dutch-American Peter van de Kamp of Sproul Observatory in Pennsylvania. In a vigil which began in the 1930s and continues even into his current retirement he searches for the tell-tale wobbles in the motion of nearby stars that reveal the presence of unseen companions.

The reason for this wobble is that, in effect, the presence of another body unbalances the movement of the main star. As in an uneven dumb-bell, the point of balance between two objects lies nearest the largest companion. And it is this point of balance that describes the orbit round the Galaxy. To demonstrate the effect, van de Kamp spins a model of a star-and-planet system round its point of balance. The star wobbles slightly in its path as the planet orbits round it. The same thing happens with the Earth and Moon, which wobble either side of their common point of balance as they orbit the Sun.

Van de Kamp's telescopic tracking of the nearby stars revealed several that were not following the straight-line path that would be expected for single stars. Instead, they followed a snaking curve. The amount and frequency of such a wobble reveals the size and orbit of the object that produces it. In one case, the wavy path was so great that the object causing it could not have been a planet. Subsequently, photographs revealed one of the smallest stars known. This discovery both vindicates the technique, and confirms that there is a range of different-sized objects graduating from planets to stars.

In 1963, van de Kamp staggered the scientific world by announcing that the nearby Barnard's star had a companion only 50 per cent larger than Jupiter. That meant that it had to be a planet. Barnard's star, named after a famous American astronomer, is the fastest-moving star in the sky, but is too faint to be seen without a telescope. Excitingly, it is the second closest star to the Sun (5.9 light years away); the closest star, α Centauri, is actually a triple-star system. If the second-closest star to the Sun actually does have planets, astronomers' ideas about the frequency of planetary systems may be spectacularly confirmed.

The story of Barnard's star is not yet over: in 1969 van de Kamp published an alternative analysis which showed that there could be two planets orbiting Barnard's star with periods of 26 and 12 years. The masses of these planets he calculated to be about 1.1 and 0.8 times Jupiter's mass. The total wobble, therefore, is a combined effect of these two planets. It does not take into account the fact that there may be other, Earth-sized planets round the star, which have too small an effect to be noticeable.

The distances of the planets' orbits are roughly equivalent to the distances from the Sun of the asteroid belt and Jupiter in our own Solar System. Astronomers have therefore begun to look upon the Barnard's star system as a scaled-down version of our own. Barnard's star itself is smaller than our Sun, and hence appears so faint.

At a conference on the origin of the Solar System held in Nice, France in 1972, astronomers David Black and Graham Suffolk of NASA'S Ames Research Center presented another analysis in which they argued that there was 'strong evidence' for the presence of three massive planets round Barnard's star. The masses of the planets were roughly 1.2, 0.6, and 0.8 times Jupiter's mass, with orbital periods of 26, 12 and 7 years. But, unlike the Solar System where all the orbits of the major planets lie nearly in a plane, Suffolk and Black found that the planetary orbits round Barnard's star are apparently tilted at large angles to each other. If true, this would present a problem to theorists.

Most astoundingly of all, in 1973 Oliver Jensen and Tadeusz Ulrych of the University of British Columbia applied a new statistical technique to analyse the observations of Barnard's star, and came up with the probability that there are five major members of the planetary system. Jensen and Ulrych confirmed the existence of a planet with a 26-year orbital period, plus one with a period they put at around 11 years. They also believed there may be planets orbiting Barnard's star once every 3.8, 2.9 and 2.4 years. The masses, they calculated, range from 0.7 to 1.6 times Jupiter's mass, and they also predicted that the orbits are highly inclined to each other. However,

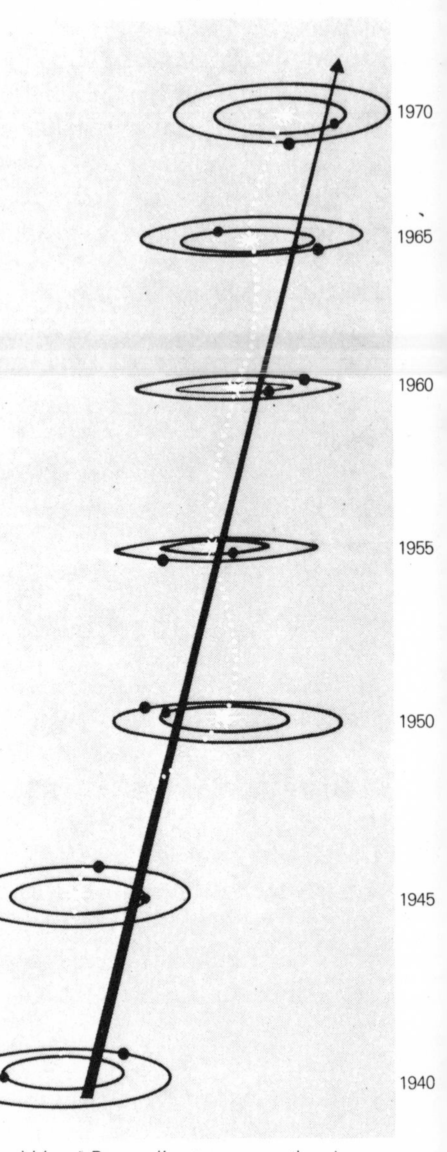

The wobble of Barnard's star across the sky, as observed since 1940, reveals the presence of at least two planets in orbit around it. Not to scale. (*After Peter van de Kamp*).

Peter van de Kamp and the 24-inch refracting telescope of Sproul Observatory, with which he has discovered possible planetary companions of other stars. (*Sproul Observatory*)

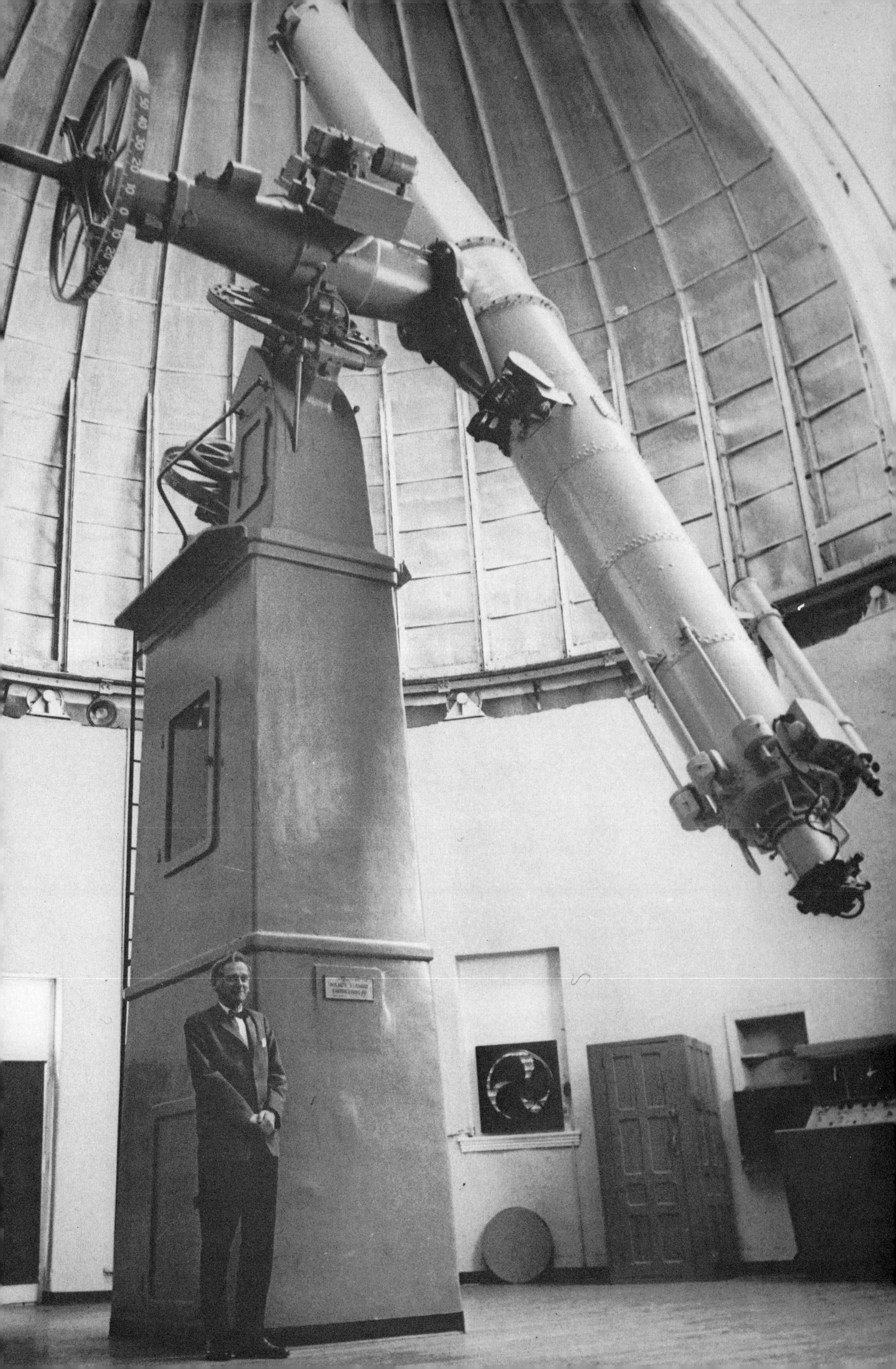

after discovering a slight error in their calculation, Jensen and Ulrych later reduced their confidence in the existence of the two planets with the shortest orbits.

For many astronomers, this is a case of over-extending limited data. And, in fact, not everyone agrees with van de Kamp's observations.

In 1973, the American astronomers George Gatewood and Heinrich Eichhorn reported an unsuccessful search for evidence of the waviness in the motion of Barnard's star on photographic plates taken at other observatories. 'We conclude, with disappointment, that our observations fail to confirm the existence of a planetary companion to Barnard's star,' they wrote in the *Astronomical Journal*, suggesting that errors in the telescope or measuring process could be responsible for van de Kamp's data.

Stung by this, van de Kamp had the Sproul plates rechecked, only to confirm his original results. And he now adds that he finds Barnard's star moving as he predicted it should if it has accompanying planets.

In private conversation van de Kamp tells me that, although Gatewood and Eichhorn have not published their data in full, he has had a chance to analyse it and finds a clear twelve year period, in accord with the orbit of one of his predicted planets.

There the matter rests, with most astronomers backing van de Kamp because of his many greater measurements—and the fact that his results have already led to the discovery of faint companion stars.

The Sproul team has noted several stars in our vicinity that have low-mass companions, but very few of these seem small enough to be true planets. In 1960 van de Kamp's colleague and successor, Sarah Lee Lippincott, published data suggesting that an object about 10 times Jupiter's size orbited the star Lalande 21185. This star is 8 light years away, and the mass of its companion, if correct, would place it on the borderline between being a planet and a star. But in 1973 George Gatewood put a question mark against this star when he failed to confirm the Sproul data from plates taken at the Allegheny Observatory.

Observers at the distance of Barnard's star would just be able to detect a wobble of the Sun caused by the planets Jupiter and Saturn. This slight effect has to be taken into account when assessing the wobble that we see of Barnard's star.

One of the nearby stars that is most like the Sun is called Epsilon Eridani. Many astronomers hold out hopes that it, and a similar star named Tau Ceti, possess solar systems very similar to our own.

Although van de Kamp has insufficient data to draw a conclusion about a possible wobble of Tau Ceti, in 1973 he presented results from plates taken between 1938 and 1972, which suggest that Epsilon Eridani has at least one planetary companion. Epsilon Eridani shows a wobble with a 25 year period, evidently caused by an object about six times Jupiter's mass. As with Barnard's star, this interpretation is open to modification as more observations become available.

Assuming that we do confirm the existence of planetary systems of other stars, how sure can we be that any of them are suitable for supporting life?

Su-Shu Huang (known to his Northwestern University colleagues as Snowshoe) has been foremost in outlining the habitable regions round stars where one would expect life to form. Huang likens a star's warming ability to a bonfire in a field on a cold night: the bigger the fire, the wider the zone of comfort. The habitable regions round a star will be where conditions are not so hot that water evaporates, nor so cold that it freezes.

Prime requisites in the search for life-bearing planets are a star of the right generation, and one which has a stable light output. As we saw in Chapter One, stars born too early in the history of the Galaxy would not have enough heavy elements from which to form planets. And an unstable star that constantly flared and faded would probably make conditions in its vicinity too variable for life. Equally, the largest and hottest stars burn out too soon. Stars which exist for less than a few thousand million years are unlikely to give life enough time to evolve to the pitch of intelligence—though, as we have seen for Earth, very simple forms of life may originate inside a thousand million years.

If we are interested in technological civilizations—that is, ones with which we can communicate—we are left with stars ranging from those that are somewhat hotter than the Sun, to those that are considerably cooler (which includes Barnard's star).

Astronomers are now finding that the tiny faint stars called red dwarfs, like Barnard's star, are vastly more abundant in the Galaxy than they had previously realized. Such stars' faintness, naturally enough, has so far made them difficult to spot. As the stable lifetime of such a star is calculated to be in the region of 100,000 million years, even the clumsiest process of evolution would have ample time to produce an advanced being. But their very dimness raises another problem: the habitable zone round them would be very narrow. The dimmest stars will warm only those planets that huddle closest to them, and it is significant that the planets of Barnard's star seem to have orbits that are correspondingly smaller than in our own Solar System.

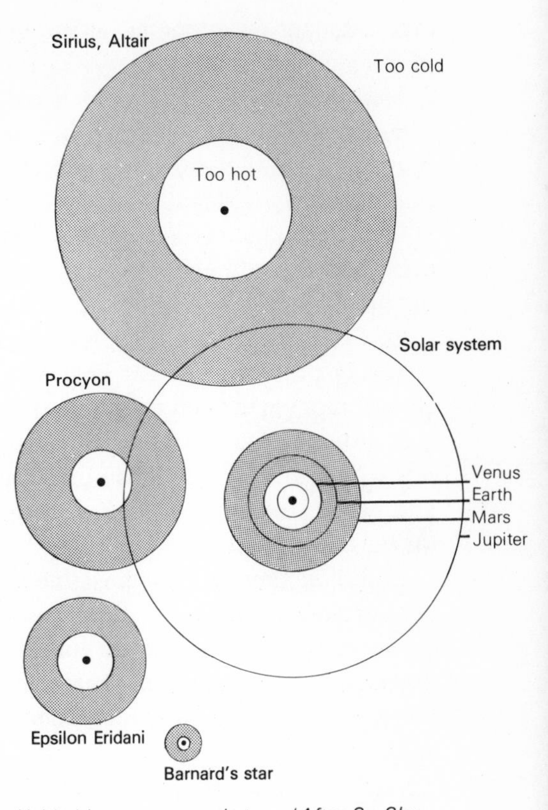

Habitable zones around stars. (*After Su-Shu Huang*)

Huang concluded on theoretical grounds that stars such as our Sun are best fitted for supporting intelligent life. The Earth, he noted, is centrally placed in the Sun's habitable zone; Venus and Mars are just within the limits. As he looked around, he found that Tau Ceti was the most like the Sun of the nearby stars, with Epsilon Eridani a close second. We shall come across these stars again in our discussions of interstellar signalling.

The former German rocket scientist Krafft Ehricke, now with the space division of North American Rockwell, in California, has also examined the effect of parent stars on the development of life round them. He argues that a star will affect the evolutionary process on planets round it, because higher radiation levels and hotter temperatures combine to increase the mutation rate. And life-span tends to be shorter in higher temperatures, so that the flow of generations is speeded up. For Ehricke, this suggests a link between star type and speed of evolution. In other words, the hotter the star, the faster the rise to advanced life-forms, and vice versa.

A star needs to be only about 50 per cent more massive than the Sun for its lifetime to be half the Sun's. Therefore, intelligent creatures evolving round such a slightly bigger and hotter star would probably be faced with its imminent demise. The pressures on them to develop interstellar travel would be correspondingly greater.

The nearest star to us of this type is Procyon, in the constellation of Canis Minor, which is 11.4 light years distant. But there are few other stars of the same type in the Sun's neighbourhood. So it is unlikely, says Ehricke, that we shall encounter travellers from such a star in search of a new home.

This is just as well. Peter Molton has imagined the slightly alarming situation that might arise if beings that evolved round a very hot star came to Earth. The peak output of their star would be at much shorter wavelengths than the Sun. Their eyes would therefore be sensitive to much shorter wavelengths, as our eyes have evolved to detect the peak of the Sun's emission.

Molton imagines: 'Earth has a visitor from a planet whose star is an ultraviolet emitter; being unable to see, he sets up his light source, blinds a few officials, kills some plants. Our reaction would be one of panic, at the very least.'

Of course, if we landed on the planet of an infra-red star and set up our own searchlight, we would produce the same effect on the hypothetical life-forms there.

Round the long-lived, cooler stars, life would proceed at a more leisurely pace. But Krafft Ehricke points out a disadvantage of the closest planets in such a system: their spin will be braked by the star's gravitational field, so that they keep one face permanently turned towards the star. The same effect has happened between the Earth and Moon, the smaller body having had its rotation trapped by the larger body's gravity.

Evolution may not be impossible on such a planet, but the conditions of permanent day or night would produce radically different organisms from those on the rotating Earth. However, Ehricke is sufficiently dubious about an organism's chances in the environment of a red-dwarf star that he places them out of bounds for the development of life.

Once again, this makes the Sun look supremely average, thus boosting the confidence of scientists who subscribe to the so-called principle of mediocrity: that we are nothing special, and that what has happened on Earth can happen in many other places.

One advantage of hotter stars is that their habitable zone is much larger than the one round our Sun. Probably there are several planets orbiting such stars that could support life. And, as the habitable zone spreads farther from the parent star, the more likely it is that a giant planet will fall within it. Stripped of their outer atmospheres, planets like Jupiter would show a rocky core if the modern theories of planetary formation are right. So life has more planets, of greater surface area, to work upon in the regions of bigger stars.

Occasionally, therefore, we can expect that round such stars civilizations will grow up on more than one planet. Ehricke surmises that these interplanetary neighbours would be vastly more space-conscious than ourselves. Looking round, Ehricke finds that there are a number of tiny, worn-out stars called white dwarfs within about 23 light years of the Sun. These could be the relics of once-proud stars that were larger than the Sun, and whose civilizations may long ago have fled. 'They may have passed by or through our solar system millions of years ago', he says.

Yet they probably did not stop, for they were in search of a home like the one they had left.

Disappointingly, Ehricke finds that the chances of there being a similar or more advanced civilization to ours in the Sun's immediate neighbourhood are very low. But that will not stop us looking for some form of life there.

However, the first planets on which we shall search for life are those in our own Solar System.

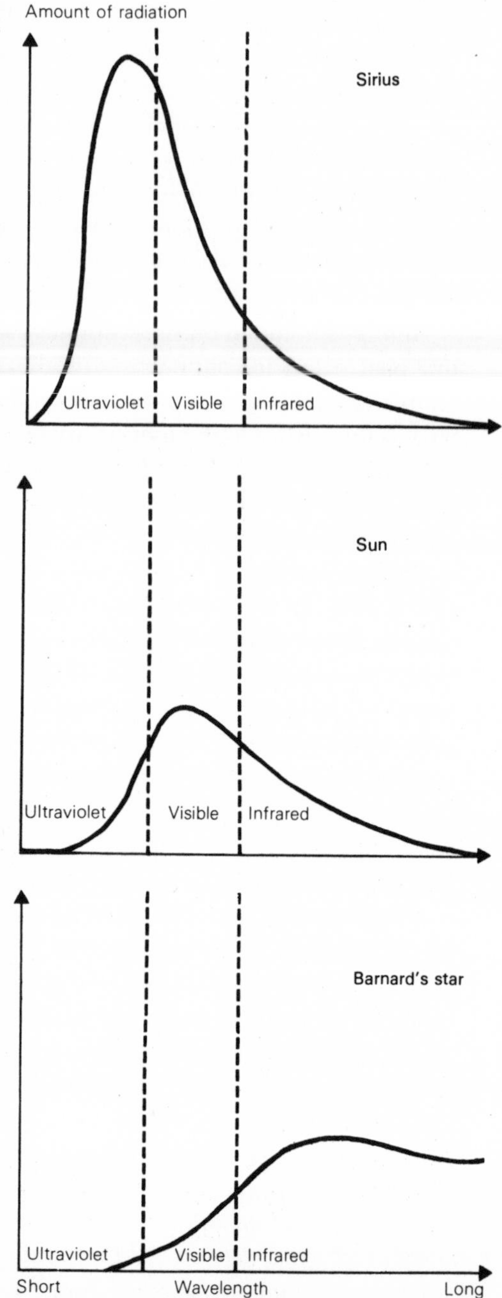

Radiation outputs of different stars, showing how their brightness peaks differ from that of the Sun.

LIFE IN THE SOLAR SYSTEM

There are nine major planets in the Sun's family, along with a good deal of rubble. The rubble takes many forms. Between the planets swoop the comets and meteorites, plus stray asteroids. As far as we can guess, all these are unlikely to harbour their own life.

Many of the planets have their own families of satellites. Some of these satellites were presumably formed from a spinning disk round their parent planet, like a smaller version of the Solar System. Others are chunks of rock that were captured as they swung past. As with the comets and asteroids, if life ever appears on these satellites, it will be because we put it there.

So where should we look for life in the Solar System? From our own limited experience, it seems that planets with some sort of atmosphere are necessary for the emergence of life—or, at least, life as we know it. But that is the only sort we need be interested in. Any other type we might be unable to detect or recognize.

Su-Shu Huang sought to calculate the likely sizes of habitable planets, assuming that they need solid surfaces and an atmosphere like our own. He concluded that their radii should lie between a few thousand and 20,000 kilometres. But, as we shall see, the contemporary view of a 'habitable' planet has very few physical boundaries.

All textbooks tell us that the closest planet to the Sun is tiny Mercury, only 50 per cent larger than our own Moon. Now, astronomer Henry Courten of Dowling College, New York, has photographed faint objects near the Sun at total eclipses since 1966. His results convince him that at least one object, possibly as much as 500 miles across, orbits about 9½ million miles from the Sun. Perhaps Courten has stumbled upon a hitherto-unknown asteroid belt. Such an asteroid belt could be a fabulous repository of valuable metals. To get them, though, we should need heat-resistant probes. No living thing could go that close to the Sun and survive.

Nevertheless, it seems that no major body orbits closer to the Sun than Mercury, which rings our star at an average distance of 36 million miles. Bathed in ultraviolet light and X rays from the Sun on one side, frozen solid by exposure to coldest space on the other, Mercury is no likely habitat for life. If there is nothing living—or even remotely biological—on the far more hospitable Moon, then Mercury must be totally sterile.

Mercury's small size and the high temperature on the side facing the Sun, mean that it can hold no appreciable atmosphere. There is a thin layer of gas round Mercury, as atoms of the solar wind are temporarily captured. But no living creature could breathe this. Being so far inside the gravitational grip of the Sun, Mercury's spin has shuddered to a near-standstill. The planet turns once on its axis in 59 Earth days, which is two-thirds of the time that it takes to orbit the Sun. One startling consequence of this is that Mercury must complete two orbits, each lasting 88 Earth days, for the Sun to go once round its sky. Equally weird effects may be at work on any small planets that orbit close to Barnard's star.

Astronomers have never been able to see Mercury's surface in any detail from Earth, and though they guessed that Mercury would look very much like our own Moon, they had to wait for space probe confirmation. They got it in 1974 when the Mariner 10 spacecraft flew past Mercury and sent back detailed pictures of its surface. The panoramas it revealed were startlingly like those of the Moon. Clearly, Mercury has been subjected to a massive bombardment by meteorites during its history. The largest and oldest marks on Mercury are clearly the scars of the mopping-up operation that went on after the planets had been formed, confirming that the first few hundred million years of the Solar System were too unsettled for life to form anywhere.

Mariner 10's instruments have also confirmed that Mercury is the densest of planets, having been formed from heavy elements in the Solar System that were not so easily spun away from the Sun. Mercury seems to have heated up inside during its early life, with lighter rocks floating to the top over a core of iron which probably occupies 80 per cent or so of the planet's interior. This process, called differentiation, seems to have operated on all the rocky planets of the Solar System including, to a lesser extent, our own Moon.

Perhaps gases once escaped from the interior of Mercury, as they still pour from volcanoes on Earth. If so, there are no signs of erosion to indicate that these gases ever formed a substantial atmosphere. In fact, Mercury gives every impression of being a totally dead planet.

Right: Mosaic of Mariner 10 photographs of Mercury, showing the planet's similarity to our Moon. Below: Simulation of Mariner 10 flying past Mercury. In its path round the Sun, Mariner 10 photographed Mercury in March 1974, September 1974, and again in March 1975. Below right: Mariner 10 being prepared for its epic mission. Before reaching Mercury, the probe also flew past and photographed Venus. (NASA).

What of Venus, often called Earth's twin because of its close similarity in size?

Venus is the brightest planet seen from Earth, and is familiar as the brilliant object referred to as the Morning or Evening Star. The brightness of Venus is due to its thick cover of white clouds, which the human eye has never pierced.

Though Venus can come closer to Earth than any other planet, its white shroud concealed the true nature of its surface. Contrasting ideas abounded among astronomers. Some suggested that Venus was covered with seas; others said the planet resembled an arid desert.

Perhaps the most charming idea suggested that under its clouds Venus resembled the steaming tropical forests of Earth several hundred million years ago, when dinosaurs roamed our planet.

But now we know that Venus is a blazing-hot and inhospitable world. Its surface rocks glow red with their high temperature, and only weak shafts of sunlight filter through the dense and unbreathable atmosphere. This picture of a hell-hole Venus has been pieced together over the past fifteen years from ground observations backed up by on-the-spot probe soundings.

Measurements from Earth first suggested that Venus has a fiercely hot surface temperature. Soviet probes confirmed this as they parachuted down through the thick atmosphere. They reported that over 90 per cent of the atmosphere of Venus is made of carbon-dioxide gas, with a trace of water vapour. Together, these gases seal in the energy of the sunlight that penetrates the clouds, in a process termed the greenhouse effect.

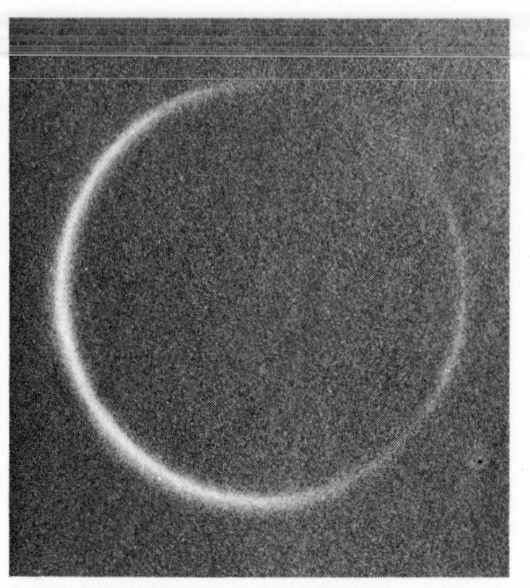

Above: Sunlight refracted through the dense atmosphere of Venus when the planet was nearly between the Earth and the Sun. (*New Mexico State University Observatory*)

The swirling clouds of Venus, photographed in detail for the first time by Mariner 10 in 1974. This computer-enhanced photograph, taken in ultraviolet light, reveals details of the atmospheric circulation of Venus. (*Philco-Ford*)

The high temperature and density of the atmosphere of Venus combine to make the surface pressure on the planet 90 times that of the Earth's atmospheric pressure. Combined with the unbreathable nature of carbon dioxide, plus the 475°C surface temperature, an astronaut who foolishly landed on Venus would be simultaneously crushed, roasted and suffocated.

The high density of the gases round Venus bend light in such a way that the land seems to slope upwards. Standing on the surface of Venus, you would appear to be at the centre of an enormous saucer.

In 1970 and 1972, the Soviet probes Venus 7 and Venus 8 parachuted to the surface of the planet. The signals from these probes confirmed that Venus has a remarkable rotation: it spins from right to left, or east to west, which is the opposite way to the axial spin of the other planets. Another outstanding fact is the slowness of this rotation. It takes 243 Earth days, which is longer than the 225 days that Venus takes to complete one orbit.

Yet visual observers report that the cloud tops of Venus swirl once round the planet, also from east to west, every four days. As Mariner 10 swept past Venus on its photo-taking mission to Mercury, it sent back the first close-up views of these gyrating cloud masses of Venus. Winds circulate strongly in the upper layers of our twin planet's atmosphere.

Radio astronomers first detected the slow, backwards rotation of Venus when they began to bounce radio-waves off its surface. More recently, by analysing the character of the reflected radio-waves, they have been able to detect large craters on Venus's surface. Instruments aboard the Soviet soft-lander probes suggested that the surface rocks of Venus were like granite, indicating that this planet too has been melted and chemically differentiated in the past. Without its atmosphere, Venus might well look very like the Moon.

Those craters may have been formed early in the Solar System's history as Venus swept up large rock chunks. Or, according to one recent idea, they may have been caused by descending fragments of a splintered moon. Such a substantial knock could have pushed Venus into its back-to-front rotation.

Why does Venus have an atmosphere so surprisingly different in composition from the Earth's? Calculations show that the amount of carbon dioxide on the Earth is roughly equivalent to the carbon dioxide in Venus's atmosphere. The vital difference is the existence on Earth of living creatures. Much of the Earth's carbon dioxide has been locked up in sediments such as limestone by the activities of sea creatures, which use carbon dioxide to make carbonate shells for their bodies. Plants, on the other hand, break down carbon dioxide, keeping the carbon for their bodies and releasing oxygen. As there is no free oxygen on Venus, we can conclude that there are no living creatures there.

Though Venus is clearly a hostile planet to life, we saw in Chapter Three how plants seeded into its atmosphere might begin to work those changes that could make its environment more like the Earth. There may well be cool, moist regions in the upper layers of the atmosphere where micro-organisms could live.

But evidence that conditions closer to the surface of Venus might not be so lethal was discovered in 1972 by Peter Molton, Joann Williams and Cyril Ponnamperuma. They sealed several types of terrestrial micro-organisms in tubes at pressures up to 20 times the Earth's atmospheric pressure, and heated them to over 100°C for a day. In most cases, significant numbers of the micro-organisms survived. 'These results may have some implications for the exploration of the atmosphere of Venus by the use of space probes,' commented the experimenters. Unless the Soviet probes were adequately sterilized, there may well be terrestrial life-forms flourishing in the atmosphere of Venus at present.

Planet Three, known to its inhabitants as Earth, holds many of the clues to life-processes in the Universe, some of which we have outlined in this book. Its habitability is the result of several coincidences: its size, its distance from the Sun, and most importantly the gases that it exhaled as the decay of radioactive atoms warmed its interior. The outward seep of gases continues today via volcanoes, but it was far more intense a few hundred million years after the Earth's formation.

We can analyse these gases in fair confidence that they have not changed much during our planet's history. The major component, often 95 per cent or so, is water vapour. That vapour condensed into rain on the ancient Earth and fell over millions of years to fill its seas.

Other components of volcanic gases are nitrogen, currently the main component of our atmosphere, and carbon dioxide, which as we have seen has been removed from the air by life-processes.

Scientists believe that this secondary atmosphere of Earth was exhaled as the first, primitive atmosphere was stripped away by the glare of the young Sun. But if the other planets had a similar history to the Earth, why are they not also dripping with water? The answer seems to be that the water molecules have been split into their component atoms of hydrogen and oxygen by the energy of sunlight. In the case of the Earth, the oxygen molecules have built up a screen of ozone high above our planet, which has filtered out further harmful ultraviolet rays.

By contrast, the higher temperature of Venus kept most of its water vapour as a gas high in the atmosphere, where it was readily broken up. If Earth had been closer to the Sun, the higher temperature would have broken up all our water. And water, biologists feel, is indeed the liquid of life.

The Earth is remarkable in having a Moon that is a large fraction of its own size. The Moon may have had subtle effects on life. The strength of its tides could bring life from the sea to the shore. And, once there, the developing intelligence of that life would be stimulated by the Moon's cyclic changes.

Eventually, that intelligent life would be compelled to reach for its Moon. A world without a convenient natural satellite might take much longer to develop spaceflight.

Our trips to the Moon have shown that its interior has melted in the past. Being a smaller body than the Earth, the Moon would generate less heat, which could also escape quickly. So it is much cooler now.

Biologists wondered if conditions early in the Moon's history might have been right for life to form, and they feared that some organisms harmful to human life might lurk below the surface. So quarantine procedures were instituted for the first returning Apollo astronauts and their cargoes of rock. But, despite careful studies, nothing remotely resembling life has been found in any of the lunar samples. The Moon is indeed sterile.

With a diameter of 2,160 miles, only about a thousand miles less than Mercury, the Moon would not be expected to hold any appreciable atmosphere this close to the heat of the Sun. Its surface has therefore been laid open to the meteorite bombardment that the Earth's denser covering has to some extent blocked.

Despite the inhospitable nature of the Moon's surface, there is

Top: The life-bearing planet Earth, showing cloud swirls in the rich atmosphere, photographed from space by the Apollo 11 astronauts.
Above: The barren and airless surface of the Moon is pitted with craters like these on the Moon's far side. The Moon's surface, like that of Mercury and also Mars, was evidently bombarded by meteorites early in the Solar System's history. (NASA)

The Apollo 12 astronauts bring back the TV camera from the robot Surveyor 3 probe, on which a terrestrial micro-organism had apparently lived for 2½ years. (NASA)

evidence that terrestrial bacteria can stand up to the conditions there. In 1969, the Apollo 12 astronauts returned from the Moon with a television camera cut from the robot Surveyor 3 probe that had automatically landed on the Moon 2½ years before. In the laboratory, scientists found a *Streptococcus* organism that had apparently survived unharmed on the camera since the Surveyor was launched from Earth. Biologists are now looking more closely at the limitations of terrestrial life. They know of micro-organisms that exist at high levels of temperature, acidity and radiation on Earth. The flexibility of terrestrial organisms is enough for them to survive some of the most unpromising environments in the Solar System.

Botanist Professor S.M. Siegel of the University of Hawaii has been in the forefront of investigations into the behaviour of terrestrial life in extreme conditions. His experiments lead him to the conclusion that living matter has a far greater range of potentially habitable environments than biologists have traditionally supposed. One interesting discovery was that plants reared in a low-oxygen environment have far greater resistance to freezing than similar plants reared in a normal Earth atmosphere. This is just the requirement for their survival on Mars. And turtles turn out to be animals that can survive a low-oxygen and high-ultraviolet environment—which has led to the suggestion that the typical Martian may be shell-backed.

Antarctica provides a sufficiently hostile environment for studying the possibility of life-forms on cold planets. A salty pond in Antarctica, still liquid at -24°C, was found to contain living bacteria and fungi that had become adapted to the conditions.

Siegel's laboratory experiments have simulated yet more forbidding conditions. Onion seeds have sprouted in liquid ammonia at -23°C, and mould spores have germinated at even lower temperatures.

Though no one is able to show that terrestrial-type organisms could survive at the surface of Venus, Siegel's experiments clearly show that they could flourish under conditions found on Mars or the outer giant planets. If we wish to practise panspermia, we could successfully sow our seed into the environment of many habitats in space.

But could planets with such hostile conditions have given birth to life-forms of their own? One place where we might hope to find an answer is Mars, called the red planet because of its distinctive colour. Evidently, the sands of Mars are tinted by a form of iron oxide. Oxygen from the Martian atmosphere has become chemically locked-up in the surface rocks by a process similar to that of rusting on Earth.

Astronomers have been able to see for many years that the air round Mars is very thin, but not until the advent of space probes could they accurately measure just how thin it is. Even at the surface of Mars, the atmosphere is no thicker than it is at a height of twenty miles above the Earth. And the majority of the gas round Mars is carbon dioxide.

Though the surface rocks of Mars can get quite warm, the surrounding air is too thin to hold this heat. So even on a summery Martian afternoon the air temperature may not get much above freezing-point; at night, the temperature plummets well below. The chilling temperatures on Mars are well shown by the polar-caps, which are made mostly of frozen carbon dioxide—known as 'dry ice' on Earth, and used as a refrigerant. The temperature of the caps is around -125°C.

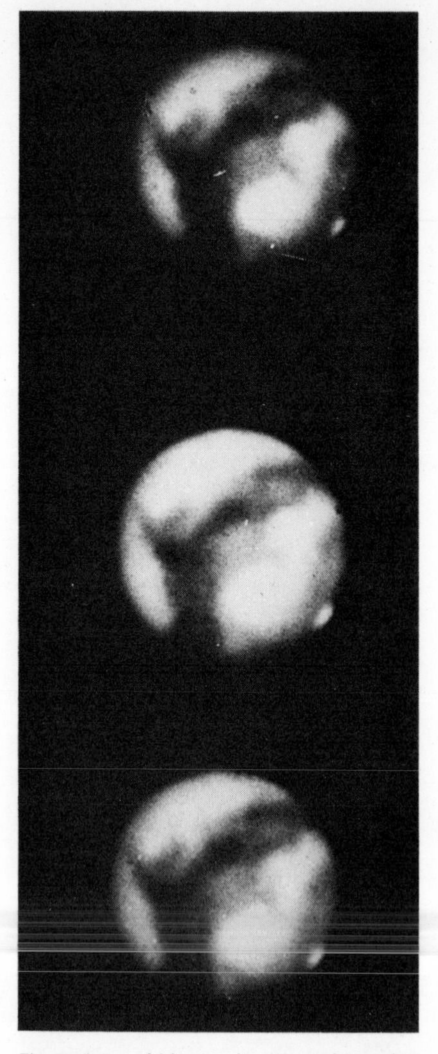

Three views of Mars as it appears in modest telescopes, showing its dark markings, bright deserts and, at the lower right, a white polar cap.

Mars would doubtless have had a more substantial atmosphere had it been a larger planet. Then, it would both have degassed more from inside, and possessed extra gravity to hold this more plentiful atmosphere. According to this view, a planet the size of Earth could be well suited for life even at the distance of Mars.

However, as we have seen, there is another school of thought which says that there already is enough gas trapped in the polar-caps of Mars to provide a substantial atmosphere, should the climate ever warm sufficiently.

Though Mars is about half the Earth's diameter, it has roughly the same land area because it is all solid surface, unlike the Earth, which is largely covered by sea.

Mars has an axial spin of just over 24 hours, so the days would seem similar in length to Earth. But, at a distance of 142 million miles from the Sun, Mars takes nearly two of our years to complete one orbit.

Most people have heard of the supposed canals of Mars, though few realize that they have always been a matter of controversy among scientists. The foremost proponent of the Martian-canal idea was Percival Lowell, a rich American who wrote books about what he imagined the inhabitants of Mars were like. He believed that they dug the canals to bring water to their crops as the Martian summer melted the polar-caps.

Astronomers agreed about certain large dark markings on Mars, and the idea that they might be vegetation was reinforced by the observation that the areas grew larger and darker as summer

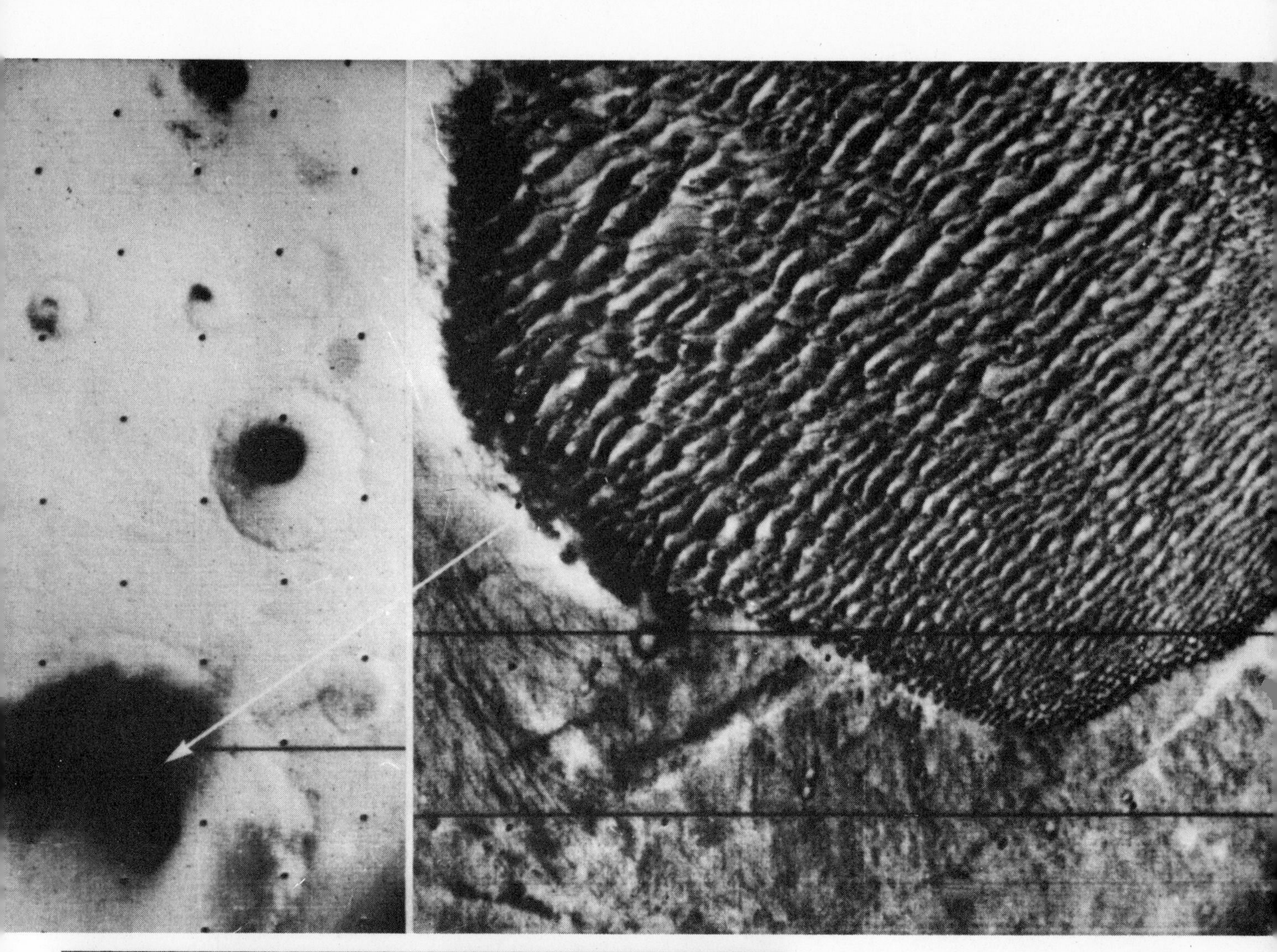

Dust dunes photographed by Mariner 9 on Mars reveal the effect of high winds at the planet's surface. (NASA)

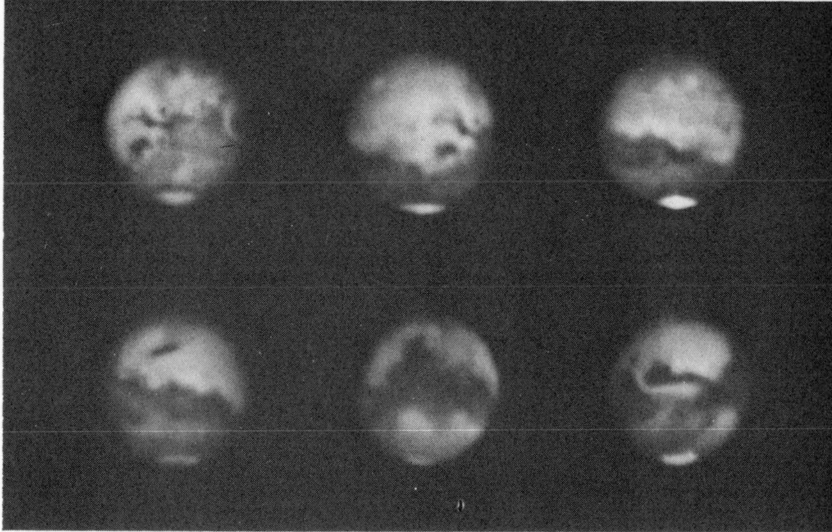

Mars photographed in red light, showing different surface markings as the planet completes one complete revolution, spinning from left to right. A bright polar cap is at the bottom. (*Lunar and Planetary Laboratory*)

progressed. Unfortunately, few astronomers could detect the straight 'canals' that Lowell drew. In fact, the lines that he drew were too thin for his telescope to have seen on Mars. Space-probe mapping of Mars has not confirmed the existence of canals. Probably Lowell was the victim of an optical illusion that caused him to join up random dots into linear features.

The variability of the confirmed dark patches remains something of a puzzle. They seem to be the result of dust on Mars being blown round by high-speed seasonal winds.

Mariner 4, in 1965, was the first space probe to take a close-up look at Mars. It showed that Mars looks very like the Moon in parts, with large impact craters. The atmosphere, though thin, has done enough to wear many of the craters down with age.

In 1969, shortly after the first Moon landing, Mariners 6 and 7 took a wider look at Mars. They found many more craters, and measured a disappointingly low water content in the Martian atmosphere. The prospects of Martian life began to look slim.

In 1971, however, Dr Norman Horowitz and his collaborators at the California Institute of Technology published results which showed that simple organic molecules could be formed at, or just under, the surface of Mars. Ultraviolet light from the Sun, they believed, could be turning the carbon, oxygen and hydrogen in the atmosphere of Mars into substances such as formaldehyde.

Also in 1971, Fraser Fanale of the Jet Propulsion Laboratory argued that conditions on Mars would have been favourable for the development of organic molecules for about the first hundred years of its existence; a time that he describes as 'a uniquely favourable opportunity' for the origin of Martian life. Fanale believes that a substantial early atmosphere would have arisen round Mars as a result of degassing from its interior. That atmosphere was lost, he says, through leakage to space, and chemical combination with surface rocks.

At the same time as Fanale published his ideas, another spacecraft on its way to Mars was to increase scientists' optimism. This was Mariner 9, which on November 13th, 1971, became the first man-made object to orbit another planet.

During one year, Mariner 9 made a photographic map of Mars, revealing amazing detail that had not been seen by the more cursory glances of its predecessors.

Nix Olympica is an enormous volcano on Mars. The lava flows surrounding it show that it was active until very recently. This oblique view was made from a globe of Mariner 9 photographs. (*Jet Propulsion Laboratory*)

The most significant discovery was the existence of young, large volcanoes. Mars, in fact, possesses the largest volcano known: Nix Olympica, which is even bigger than the volcanic Hawaiian islands on Earth.

Also visible on Mars are winding channels that look like dried-up rivers. These are too small, and in the wrong places, to have been the canals that Lowell claimed he saw. The scientists concluded that, in recent geological time, the venting of gases from the Martian volcanoes had produced a temporary atmosphere with torrential rains.

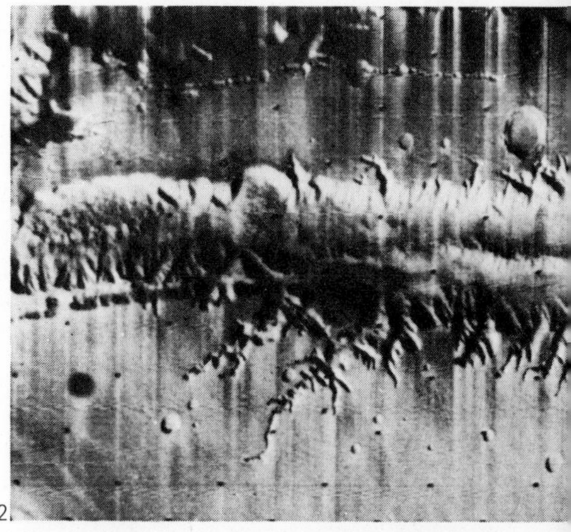

1 Mariner 9 photographed what appear to be dried-up river valleys on Mars, evidence that the climate was once less arid than it is today. The surface is also covered with craters caused by the impact of meteorites, and is streaked by wind-blown dust. (NASA)

2 A Martian 'grand canyon', evidently formed by geological faulting in the crust of Mars and further eroded by wind-blown dust. (NASA)

Perhaps today there are warm, moist cracks near those volcanoes where simple organisms can live and grow. Certainly, Mars from time to time in the past seems to have possessed the same conditions that scientists believe led to life on Earth: a primitive atmosphere, volcanic heat and downpouring rain. The question that remains is whether Mars kept these conditions long enough for life to start.

The instruments aboard two Viking spacecraft are designed to find this out. After an eleven month journey, the Vikings are scheduled to land on Mars in July and August 1976. They are identical craft, made in two parts: the orbiter, which remains circling Mars; and the lander, which drops to the surface to make on-the-spot measurements.

The Mariner 9 photographs were used to select landing-sites for Viking. They had to be flat in order to give easy touchdown (Viking has a mere nine inch ground clearance), and in lowland areas, to give the braking parachutes maximum drag. In addition, the biologists wanted sites that were wet and warm — the most likely conditions in which one would find life.

One site, called Chryse, is at the end of a long Martian canyon down which scientists believe water flowed during the wet times on Mars. Interesting sediments are expected here.

The other major site is Cydonia, an area where the scientists are hopeful that liquid water still exists on Mars.

When the Viking craft go into Martian orbit, their on-board television cameras will take close-up looks at the selected sites to confirm their suitability. Also, other instruments will scan the planet for pockets of warmth and moisture. If the intended targets seem unsuitable, the landers can be sent to different locations on the surface. The orbiter sections of the Viking craft will continue the reconnaissance of the previous Mariner probes, though in more detail; features as small as 50 metres across will be visible, which is twice as much as Mariner could discern.

The lander, sterilized to prevent contamination of Mars, is contained in an aerodynamic shell, which helps to brake its fall through the Martian atmosphere and also shields it from the frictional heat of atmospheric entry. After the lander's speed of descent has slowed somewhat, parachutes open for its final fall to the surface, with small rockets providing a final cushioning effect.

The lander can communicate either directly with Earth or via the orbiter. During the descent it will have been sensing the atmosphere of Mars, and it will continue to study meteorology once on the ground. Other experiments will listen for Marsquakes, and small magnets will test the magnetism of the Mars soil.

But the major interest lies in the experiments that test for life on Mars.

Describing the Viking programme in the planetary magazine *Icarus*, project scientists Gerald Soffen and Thomas Young of NASA's Langley Research Center said:

> The overall goal is to explore Mars as broadly as possible, gathering data on the planet's atmosphere and surface. The particular emphasis placed on the biological aspects of the investigation reflects the intense scientific interest in this question, rather than any weighted expectation of obtaining a positive result.

When Viking arrives for its three-month study, summer will be under way in the northern hemisphere of Mars. If there is life on Mars, it should then be blooming.

An artist's impression of the Viking lander on the surface of Mars. The soil sampler is seen digging trenches in the surface. (NASA)

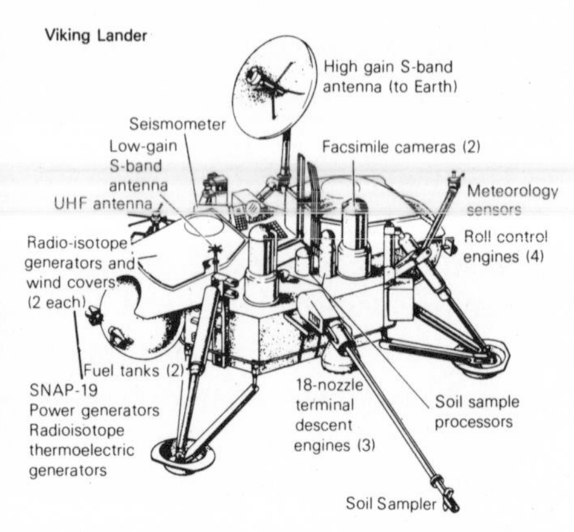

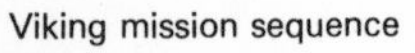
Viking mission sequence

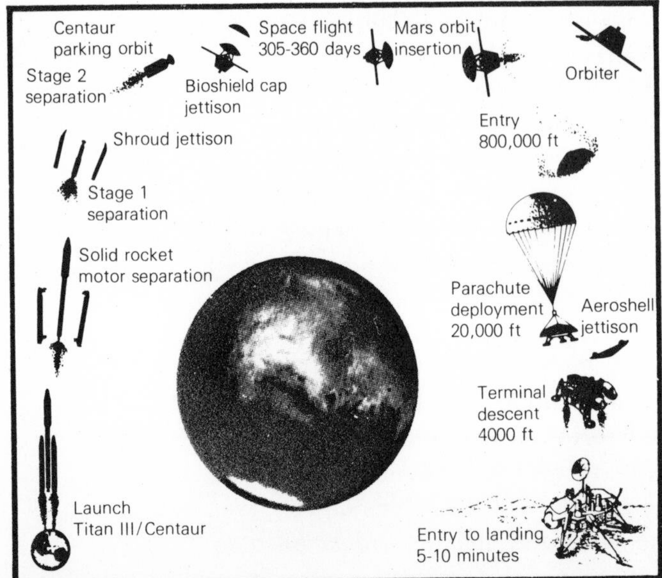

How the Vikings will reach Mars. One part of the craft remains in orbit while the lander touches down.

Two cameras on board the lander will swing round to take 360° panoramas of the Martian surface. Combining these pictures will give a stereo effect to aid scientists in mapping the lander's surroundings. Near the footpads, the cameras will resolve details as small as one-tenth of an inch across.

Among the goals that the scientists directing the television experiments list are a search for living or fossil life-forms, observation of any burrows, tracks or trails, and a search for 'technological artifacts'—in other words, constructions by Martians. The team say that the most exciting outcome would be the recognition of forms that are conclusively living. If a live creature walks past the camera, it will be detected—even if its biochemistry is totally alien.

The television cameras will also give the first Martian's-eye view of the strange moons of Mars as they spin overhead. These romantic bodies, once suspected of being abandoned space stations, were shown to be captured asteroids by the cameras on Mariner 9.

As the season progresses, the lander's cameras will look for changes in the soil such as would be caused by growing vegetation. Overhead, the instruments of the orbiter will keep watch to confirm the results from the lander. Perhaps, once the experiments with the lander are complete, the orbiter's path can be dropped lower so that it will take pictures of Mars with as much detail as the best probe photographs of the Moon.

Viking's cameras will be able to survey its ten-foot-long scoop as it digs into the Martian topsoil. Soil samples will be hauled into the lander probe for analysis.

A series of samples will be analysed by a device called a mass spectrometer—an instrument using the same technique as was employed in searching for organic molecules in lunar samples and meteorites. The mass spectrometer sorts out molecules according to their weight, by which they can then be identified.

Any organic compounds present will be turned to gas as the soil is heated in a tiny oven. The equipment is designed to detect organic molecules of any biochemical make-up. As the experimenters say, 'It is not at all clear how far living organisms, if they exist on Mars, must resemble those with which we are familiar.'

Even if there is no life on Mars, the experiment will be able to show what stage the chemical build-up of molecules has reached. 'The nature of the organic material present on Mars is very relevant both to the problem of the origin of life on Earth and the evidence for life elsewhere in the Universe,' say the molecular analysis team, led by Klaus Biemann of the Massachusetts Institute of Technology.

The equipment will also 'sniff' the Martian atmosphere to determine accurately its composition—slight changes during a season could reveal the presence of life; and the exact composition of the rocks will be analysed by additional equipment.

If some form of life has arisen on Mars, a team of biologists, led by Harold Klein of NASA's Ames Research Center, hope to find it. They have cast their net wide in an attempt to detect at least one of the characteristics of Martian micro-organisms.

Soil from the Viking's lander scoop will be shaken into a series of cups, each of which will be subjected to a different test. For the first test, the biologists will be looking for evidence of photosynthesis—that is, the uptake of carbon from the gases of the atmosphere. 'If carbon-based life exists on Mars, it is hard to imagine its being isolated from the large reservoir of carbon compounds in the atmosphere,' comment experimenters Norman Horowitz, Jerry Hubbard and George Hobby of the California Institute of Technology.

They are using carbon dioxide (CO_2) and carbon monoxide (CO) gas that has been 'labelled' with radioactive carbon. Equipment on board the Viking lander will be able to measure whether any of this carbon is taken up by the Martian soil sample, which would happen if there were living organisms in it.

The value of their experiment is that it will test for life under Martian conditions; CO_2 and CO gas are contents of the Martian atmosphere. But another experiment will be looking for signs of active life by dampening the soil with a solution of organic nutrients. These are labelled with radioactive carbon and sulphur. If the nutrient mixture encourages Martian micro-life to grow, the experimenters will be able to detect it since any gas given off will contain radioactive atoms.

This experiment was originally called 'Gulliver', a humorous reflection on its task to look for little men. 'A single cell should produce a response,' is the hopeful remark of its originator, Gilbert Levin, who has developed the idea over more than a dozen years.

A third experiment will look for changes in the atmospheric content of several gases that may be affected by the presence of life. A sample of the Martian atmosphere will be trapped in the detector along with the soil sample. The sample will be incubated in water containing nutrients such as amino acids, salts and vitamins, while a detector will look for changes in the amounts of hydrogen, nitrogen, oxygen, methane and CO_2.

These gases are likely to be either consumed or produced by Martian organisms. Changes in the gas composition of the trapped Martian atmosphere should allow conclusions to be made about the presence of life on Mars, says experimenter Vance Oyama of NASA's Ames Research Center.

If Carl Sagan's Long-Winter model of Mars is correct (page 57) there may be spores and vegetative forms of life waiting in hibernation for the arrival of Martian spring. On Earth, such life-forms are known to survive for centuries, at least. So there is no reason in principle why they could not survive on Mars for thousands of years.

It was Carl Sagan's wife Linda who suggested that the correct recipe for detecting Martian biology is 'add water'. In simulating the arrival of Martian spring, the latter experiment will do just that.

The two Vikings will not be the first to land on Mars. In late 1971 the Soviet Union ejected landers from their Mars 2 and Mars 3 craft. But both failed to return information. A fleet of four probes in 1974 was equally unsuccessful. Three of them sped straight past Mars without going into orbit. Two ejected landers, one of which missed the planet while the other failed before reaching the surface.

Probably men will eventually follow the robot probes to Mars, but not until the turn of the century. And if there is life on Mars, we shall have to take particular care to avoid contaminating our planet with what could turn out to be a plague of alien organisms. This same danger is inherent in bringing back any samples by remote control.

Beyond Mars lies the belt of the asteroids, broken hunks of rock like the debris from a building-site. From the asteroids may come many of the molecule-bearing meteorites that plunge to Earth. Another possible source for meteorites is from the heads of comets.

As the largest asteroid is but a few hundred miles across, they probably hold no more chemical clues to the origin of life than do meteorites.

The planet that makes the biologists' eyes gleam is Jupiter, lurking

Jupiter photographed through the 200-inch Mount Palomar telescope. The dark oval object is the Red Spot. Also visible is the moon Ganymede, casting its shadow on Jupiter. (*Hale Observatories*)

342 million miles beyond Mars. Jupiter is the giant planet. All the other planets rolled into one would not make up half of Jupiter.

Anyone who lived on Jupiter would be entitled to consider themselves the true kings of the Solar System. By comparison, the inhabitants of Earth live on a tiny and unimportant lump of rock.

Jupiter is made of gas; it is eleven times the Earth's width, and a third of a million Earth-type bodies would be needed to outweigh it. Jupiter is made of the same gases as the Sun. In fact, if Jupiter were several times bigger it would qualify as a small star.

In looking at Jupiter we are seeing the same type of atmosphere as surrounded the Earth shortly after its formation. Jupiter's greater gravity means that even the lightest gases cannot escape. So it has retained the envelope of gases that it grabbed from the nebula round the Sun.

Jupiter is a fossilized version of the early Earth. It is surrounded by colourful clouds, drawn out into long belts by the swiftness of its rotation—under ten hours for one complete spin, the fastest in the Solar System.

The white, wispy clouds are made of ammonia frozen into crystals by the sub-zero temperatures. Below these are darker, richly coloured regions of ammonia and water vapour which stand out clearly in Earth-based telescopes. There is also a thin surrounding atmosphere of hydrogen and methane gas, while hydrogen sulphide probably exists in Jupiter's lower layers.

'The very coloration of Jupiter may be a sign of a large-scale formation of complex molecules of carbon, nitrogen, oxygen, and sulphur,' says Carl Sagan.

In 1959, Sagan and Stanley Miller made the first experiment on a Jupiter-type mixture of gases by passing an electric spark through it. The results were similar to those obtained from simulations of the pre-

Cloud top
Ammonia crystals
Ammonia droplets
Ammonia vapour
Ice crystals
Water droplets
water vapour
Liquid hydrogen and/or solid hydrogen
Radius = 0.8
Transition to metallic hydrogen
Radius = ?
Metallic hydrogen
Transition to dense core of metals and silicates (perhaps 10 Earth masses)

A cross-section through the clouds of Jupiter, showing the different chemicals that are believed to exist at each layer. Jupiter is such an enormous planet that its vast pull of gravity compresses hydrogen first into a liquid, then a solid, and finally into a super-solid in which hydrogen takes on some of the properties of a metal such as conducting electricity. This conducting core of metallic hydrogen is believed to be responsible for Jupiter's enormous magnetic field. Perhaps some rocky material like that which makes up the Earth is also present at the core of Jupiter.

biotic Earth. The same molecule-forming agents as are on Earth are available on Jupiter. The planet is bombarded with ultraviolet light from the Sun, meteorites from space, charged particles from its surrounding radiation-belts, and there is probably thunder and lightning deep within its clouds.

Surveying the mechanisms and the material available for pre-biotic synthesis, Sagan says; 'Very large quantities of organic molecules must exist on Jupiter today.'

The temperature at the top of the clouds is not very hospitable—around -125°C—but nearer the planet it warms up considerably. There are regions below the visible clouds where temperatures are believed to be like they are on Earth—around 20°C. Though the greater gravity of Jupiter makes the pressure there as great as on the surface of Venus, that in itself is no block to the existence of life.

But it does mean that the experimenters must take the conditions carefully into account when trying to obtain complex molecules from a Jupiter-like mixture of gases. In a series of computer simulations of the Jovian environment, Sagan and his colleagues attempted to predict what molecules might exist. They found that a range of hydrocarbon compounds seemed likely to be formed, some of them with the coloured tones that observers see on Jupiter. Yellow, for example, they attributed to asphalt.

In 1969 Fritz Woeller and Cyril Ponnamperuma announced that they had studied possible reactions in Jupiter's atmosphere by sparking a mixture of ammonia and methane. They produced a deep-red substance that they attributed to molecules of hydrogen cyanide (HCN) and cyanogen (C_2N_2). A brownish-coloured substance obtained by Sagan and Bishun Khare in similar experiments is attributed to chains of sulphur atoms.

Either of these substances could explain the colour of a mysterious marking on Jupiter called the Great Red Spot. It is indeed great—a line of three Earths could be placed inside it. The Red Spot is celebrated because it is the only permanent feature on Jupiter. All the other clouds continually swirl and change like weather systems on Earth. The spot's exact colour changes with time, but is usually pink or brownish. There have been several theories to explain its permanence.

Almost certainly it is not a solid object, though for a time astronomers supposed that it might be like a raft of material floating in Jupiter's clouds. One popular theory supposed it was a column of gas swirling round some obstruction hidden in the clouds below. The Red Spot would therefore act like an eye into the region below Jupiter's clouds.

But in recent years it has become clear that the spot is an elevated region of cloud, formed as hot gases rise from deep within Jupiter, and then cool and spread out like the anvil-shaped cloud over a thunderstorm on Earth.

The reason for the Red Spot's permanence is that Jupiter is giving off heat of its own; it emits about $2\frac{1}{2}$ times as much energy as it receives from the Sun. There is thus a continual outflow of heat from the planet, and the Red Spot acts like a kind of meteorological safety-valve. Jupiter's heat excess is believed to be due to the fact that the planet is shrinking by an unobservable 1 millimetre a year. The same process was once proposed by Lord Kelvin to account for the Sun's energy.

Presumably, the rising gas column of the Red Spot stirs up complex organic compounds from below the clouds, and these cause its characteristic tinge.

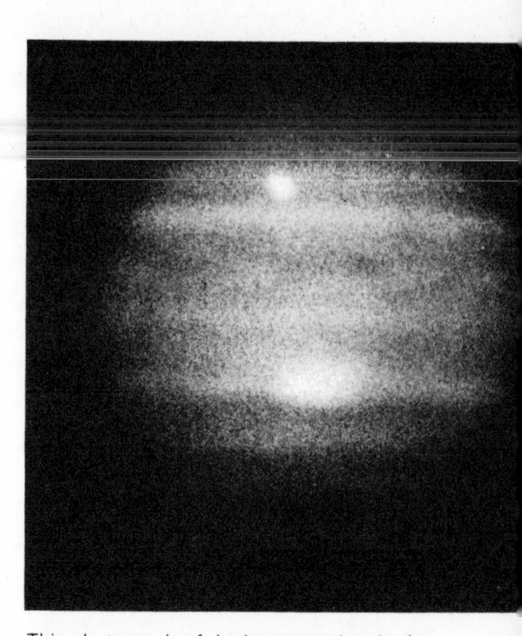

This photograph of Jupiter was taken in the wavelength at which methane absorbs light. There is methane gas in the atmosphere of Jupiter, which blocks light reflected from the clouds below. In this picture, therefore, the brightest areas are the highest, because the light from them has had to pass through less methane. The bright spot on the top half of Jupiter is caused by its moon Io passing in front of the planet. But the white oval in the southern hemisphere is the Red Spot, showing that it is higher than any of the other clouds on Jupiter. (*Catalina Observatory*).

The warm interior of Jupiter means that liquid water and liquid ammonia might be expected below its clouds. There may therefore be seas of a kind on Jupiter. And Isaac Asimov has mused: 'Think of the fishing!'

It seems likely that the complex organic molecules on Jupiter might fall to the lower layers, where they would encounter a suitable solvent. Jupiter, and similar giant gaseous planets could be likely homes for life that uses liquid ammonia.

Experimenters have found that, as on Earth, amino acids are easily formed from Jupiter's atmospheric gases. But there are problems in the assembly of these individual structural units into the very large protein molecules of life, for there are no dry shores for them to be washed on to by the gentle swell of the tides. Perhaps the convective gyres stirred by Jupiter's inner heat help the gradual assembly of the molecules of life—though it is an optimistic supposition.

The organic chemistry of Jupiter is a rapidly developing and complex field. In their laboratory at the University of Maryland, Cyril Ponnamperuma and Peter Molton have experimented widely with possible synthesis of organic compounds in Jupiter's various layers, for the first time simulating the extreme pressures to be found under Jupiter's high gravity. They conclude that pre-biotic synthesis on Jupiter may have reached an advanced state; and they do not deny the possibility that forms of micro-organism similar to those on Earth may have arisen.

Life in the clouds of Jupiter, if it exists, might have developed into floating gas bags like those in this artist's impression. From the Ian Ridpath/David Hardy filmstrip *Exploring the Planets. (Hulton Educational Publications)*

These organisms would have a ready food source in the organic molecules raining down from the higher layers on Jupiter. Their metabolism might even break the molecules down again to methane, ammonia and water to give themselves energy. Minerals for the organisms would come from the micrometeorites that Jupiter sweeps up from space. Simple organisms, not breathing oxygen, might therefore float among the clouds of Jupiter. Possibly fish-type creatures, suspended like gas bags, have arisen—but that is pure speculation.

Ponnamperuma and Molton have subjected terrestrial micro-organisms to the sort of conditions that would be expected at various levels throughout Jupiter's cloud layers. The micro-organisms—three types of bacteria and one of yeast—had to withstand pressures between 0.1 and 120 atmospheres and temperatures from 30°C to -200°C for a day. In nearly all cases, a significant proportion of them survived.

A major advance in the understanding of possible organisms that could exist on Jupiter came in late 1973 with the discovery by NASA scientists Paul Dean and Kenneth Souza of a terrestrial bacterium living in a highly alkaline spring in California. Dean and Souza, of the Ames Research Center, found that this rod-shaped bacterium thrives in a solution of sodium hydroxide ten times stronger than that which other terrestrial bacteria can stand. Although sodium hydroxide is not expected under the clouds of Jupiter, ammonium hydroxide, formed by the dissolving of ammonia in water, is believed to occur. So an ammonia-based organism might have no trouble in surviving the highly alkaline solution of ammonium hydroxide. Dean and Souza are now looking for Earth bacteria with such ammonia tolerance, and they expect to find them because, as they say, there was much more ammonia in the Earth's atmosphere when life first formed. Ammonia-tolerant bacteria may live in areas of high ammonia concentration on Earth.

Clearly, any organism that arose in Jupiter's environment would have no trouble surviving. But this does indicate that probes going to Jupiter must be carefully sterilized to avoid contamination.

Jupiter has an impressive retinue of thirteen satellites, some of which were probably formed round it in a small-scale imitation of the growth of planets round the Sun. The second closest moon, Io, orbits at the inner edge of Jupiter's strong radiation belts. Io, which is bigger than our own Moon, seems to have a thin atmosphere, and it is intriguing because it is noticeably orange in colour—similar, in fact, to the Red Spot. This suggests that organic synthesis is important on at least some of Jupiter's moons.

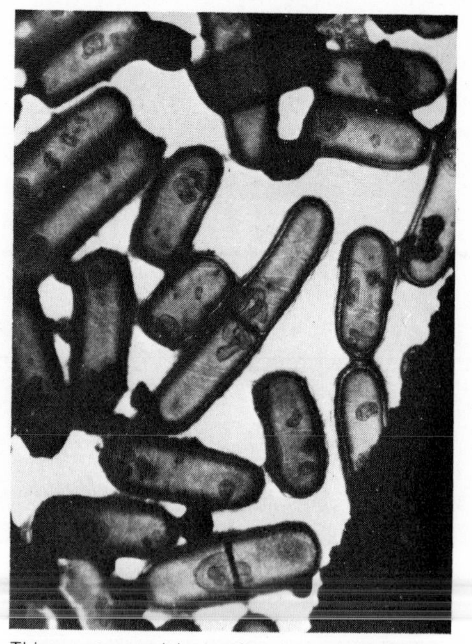

This rare terrestrial organism can survive and grow in an environment with similarities to that believed to exist on Jupiter. The bacteria swim, grow, and reproduce in an environment as alkaline as Jupiter is believed to be. The organisms, photographed in a group by an electron microscope, are seen reproducing by dividing into two. (NASA)

In December 1973 the American Pioneer 10 spacecraft flew past Jupiter, largely confirming what had been seen or surmised about the planet and its environment by Earth-based astronomers. A major haul from the probe was a series of superb photographs, showing Jupiter's complex cloud-swirls in far greater detail than can be seen from Earth. These are helping meteorologists to understand more about the bafflingly changeable wind systems on Jupiter.

It was a relief that Pioneer survived its passage through the asteroid belt without suffering any major impacts from flying particles. All evidence suggests that the meteorite hazard once feared as a potential killer for space travellers has been dramatically over-estimated.

A second probe of similar intent, Pioneer 11, also survived its journey to Jupiter, passing the giant planet's polar region in December 1974 and then swinging on for a close-up look at Saturn.

Even though they travelled at the highest speeds of any space probes, the Pioneers took twenty-one months to reach their Jupiter target. Pioneer 11 is not expected to reach Saturn until September 1979—but it may well have ceased working by then, for this is far beyond its designed lifetime.

In 1977 two more advanced spacecraft will follow the pathfinding Pioneers to examine Jupiter and then fly on to Saturn. These are substitutes for the abandoned Grand Tour missions, which were planned to examine all the outer planets.

Saturn, the ringed planet, is a beautiful enigma. Why should it alone have developed a major system of rings? The popular theory once was that the rings are the shattered remains of a former moon; but most astronomers now believe that they represent the building-blocks of a satellite that never formed. Possibly, thinner rings of material surround the other major planets. Answering the problem of Saturn's rings will aid astronomers in their understanding of the formation of the moons of the major planets; and also, by analogy, the formation of the planets and asteroids.

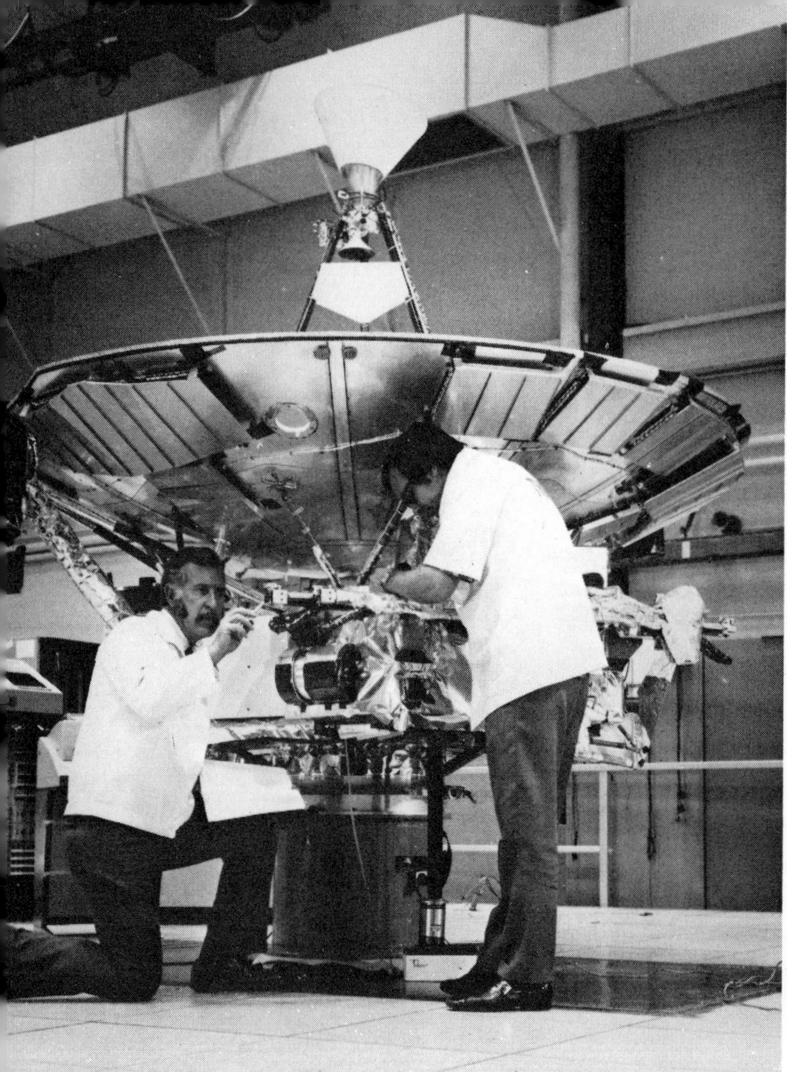

1

2

4

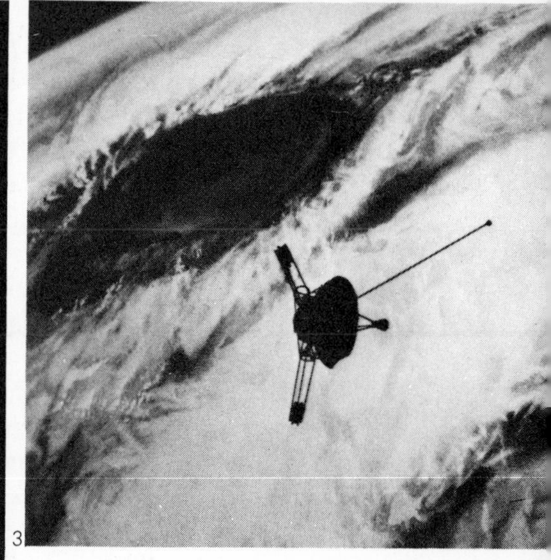

3

1,2 Pioneer spacecraft receives final inspection, and is fitted into the nosecone of the rocket that will launch it towards Jupiter. (NASA)

3 Artist's impression of Pioneer flying past the red spot on Jupiter. (NASA)

4 Jupiter photographed in close-up by Pioneer 10 on December 2nd, 1973. Cloud swirls caused by the planet's high speed of rotation are clearly visible, as well as the shadow of the moon Io. (NASA)

Saturn is another giant planet made mostly of gas. Its globe is about 75,000 miles wide, though the rings stretch almost 180,000 miles from side to side. The rings are not as solid as they look in photographs. They are made of many lumps about the size of bricks, probably coated with frozen gas. These orbit the planet like tiny moons.

Saturn spins once in 10¼ hours. As on Jupiter, its clouds are pulled out into colourful bands, but the lower temperature on Saturn means that they are less prominent and not so changeable. Probably many hydrocarbon compounds are being formed in the clouds of Saturn. One amazing fact about the planet is that its overall density is about the same as that of oil—and so, given a large enough ocean, Saturn would float on it.

A view of Saturn through a large telescope. (*Hale Observatories*)

However, this is only the *average* density. It does not have to have that value all the way through. If the planetary-formation theories currently in vogue are correct, the gaseous envelopes of Jupiter and Saturn must have collected round a solid, dense core.

As with Jupiter, Saturn is believed to have a slight heat excess that indicates it may still be shrinking. Despite its frigid exterior, Saturn's interior must be warm enough for terrestrial-type life at some deep layer in its clouds.

Saturn is effectively a scaled-down version of Jupiter, but its ten-member satellite system has the distinction of including the most intriguing moon in the solar system. This is Titan, at three thousand miles diameter slightly smaller than Jupiter's largest moon but comfortably bigger than our own Moon. Observations show that Titan has a substantial atmosphere, which astronomers are still trying to analyse in detail. This atmospheric shroud seems to keep Titan warm by the greenhouse effect—a fact that suggests to the most optimistic exobiologists that even the satellites of the Solar System are not out of the question as abodes for life.

Farther into the darkness lie the planets Uranus and Neptune. Their distance means that they have not been studied in detail. But they are cold—100° above absolute zero at the top of their cloud decks. They are similar in size—around thirty thousand miles diameter—and apparently have much the same composition. Their gaseous atmospheres appear green because of the substantial amounts of methane they contain. Ammonia probably exists lower down in the clouds.

All these gas giants may have conditions in their interiors that are suitable for floating micro-organisms, perhaps using ammonia as a solvent. But it would be optimistic to suppose that anything more than simple organic compounds will ever be found on the outermost planets.

In any case, we shall not be going there to find out for a long time.

At the edge of the Solar System—or almost—is the stray object called Pluto. Once the objective of a major search, Pluto has turned out to be very much of an interloper. It seems to be very similar to Mars in size and nature, except that it is far colder. From Pluto, the Sun's size and brilliance are so diminished that our star appears as little more than an intense point.

Most astronomers now assume that Pluto is an escaped satellite of Neptune. Pluto's orbit can bring it closer to the Sun than Neptune. And this will happen between 1979 and 1999, when Neptune will mark the edge of the known Solar System.

Astronomers have looked in vain for planets farther out than Pluto. If anything exists in the blackness beyond the known planets it is probably a cloud of comets, made of the ice and dust swept to the very edge of the Sun's gravitational grip.

And after that there is nothing until we reach the stars.

7

COMMUNICATING WITH THE STARS

Men have long strained to catch the sound of a voice from the stars. The radio pioneer Nikola Tesla believed that he had observed electrical disturbances caused by alien intelligence in 1899. 'The feeling is constantly growing on me', he wrote, 'that I had been the first to hear the greeting of one planet to another.'

It is more likely he was detecting natural signals from the Earth's magnetic field. But altogether more fascinating emissions were picked up by Marconi's telegraphists during the early days of transatlantic signalling. These were Morse letters, of much lower frequency than that used by the Marconi company.

Perhaps these were terrestrial transmissions that had been trapped in the ionosphere and then released at a lower frequency. Marconi, though, speculated that they were messages from Mars, which led to attempts to listen for Martian transmissions during the close approach of Mars in 1924.

Less-technological schemes for interplanetary contact had been proposed by several scientists (and non-scientists), such as marking out large geometrical figures on the Earth, making giant bonfires in recognizable patterns, and flashing sunlight from mirrors in a kind of cosmic heliograph.

However, with the realization that technological civilization, if it existed, would be found a lot farther away than on Mars, it became clear that radio offered the best available method of communicating across space.

Radio astronomy was effectively born in 1931 when a communications engineer in the United States, Karl Jansky, built a large aerial to investigate the origin of interference on long-distance radio telephony. He found that one noise source seemed to lie towards the centre of our Galaxy—an area where dense star clouds lie. Jansky's work was followed up by an enthusiastic radio amateur named Grote Reber, then of Chicago but now working as a professional radio astronomer in Australia.

Only after the war did radio astronomy blossom, as a result of the technological advances that the development of radar had brought about. By coincidence, the high-intensity microwave beams used for radar also served as our first unmistakable, although unintentional, messages to the stars.

In 1951 radio astronomers discovered natural radio emissions from hydrogen in space at a wavelength of 21 centimetres. Because of hydrogen's abundance, this has become one of the wavelengths most cherished by radio astronomers, and is jealously protected from the interference of other radio users.

To two physicists at Cornell University, Giuseppe Cocconi and Philip Morrison, the 21-centimetre line of hydrogen provided an obvious signalling-band for interstellar communication. In one of the seminal papers in this field, published in *Nature* on September 19th, 1959, Cocconi and Morrison proposed listening on 21 centimetres wavelength for intelligent signals from space. They foresaw that the signal might come in one-second pulses, perhaps continuing for years before repeating itself. 'Possibly it will contain different types of signals alternating throughout the years,' they said. And, in a memorable sentence, they concluded, 'The probability of success is difficult to estimate; but if we never search, the chance of success is zero.'

At the same time, the radio astronomer Frank Drake, who was familiar with the arguments for the possibility of life in space, was preparing a listening programme with the 85-foot radio telescope of the National Radio Astronomy Observatory at Green Bank. This was Project Ozma, named whimsically after the mythical land of Oz. The targets Drake chose for his celestial eavesdropping were the stars selected by Su-Shu Huang as the most like the Sun of the nearby stars: Tau Ceti and Epsilon Eridani.

Project Ozma got under way in the spring of 1960 to mixed feelings in the scientific community. Observing in the 21-centimetre region of the radio spectrum, Drake periodically switched his receiver between the two target stars and assiduously recorded the results. He had calculated that any signal being beamed to us at a reasonable strength would easily be picked up by his equipment.

Despite three months of listening, no detectable signals from another civilization entered Frank Drake's radio telescope. Drake confessed that he was not surprised by the negative result—another civilization within 12 light years would have been celestial overpopulation—but he regarded the effort as worthwhile.

Project Ozma lapsed, but was not forgotten. Periodic scares have since run through the scientific community when radio astronomers have discovered sources that appeared to be varying regularly. But in each case—such as the discovery of pulsars in 1967, which led to only half-joking talk of 'little green men'—the signals have proved to be of natural origin.

Though scientists in general are much more prepared now to accept the idea of interstellar contact than they were at the time of Project Ozma, radio telescopes have usually been too busy with their regular astronomical duties to follow up Drake's pioneering study. It came as a major surprise, therefore, when delegates at the Byurakan conference on extraterrestrial intelligence were told by Soviet astronomer Vsevolod Troitsky of star-listening sessions that had been conducted at Gorky State University.

The Gorky astronomers developed a receiver able to operate at 21 and 30 centimetres, linked to a 15-metre-wide radio dish. In their listening programme, the Soviet radio astronomers examined the region round 30 centimetres wavelength. They aimed their aerial at 11 Sun-type stars out to a distance of about 60 light years, including Tau Ceti and Epsilon Eridani, and also scanned the Andromeda galaxy in case a super-civilization of the type envisaged by Nikolai Kardashev should be transmitting there.

Each star was studied at least twice, and one as much as ten times. According to a paper published by Troitsky and three colleagues at Gorky in the May-June 1971 issue of the Soviet *Astronomicheskii Zhurnal*, listening took place in October and December 1968 and February 1969, each object being observed for 10 minutes at a time. But, within the limit of sensitivity of the receiver, said Troitsky, no emissions were recorded. Unofficial reports now suggest that the Soviet researchers have extended their work to a total of 100 stars.

Troitsky announced at Byurakan that his group had gone on to search for sporadic emissions, in the form of pulses, that could result from the activities of extraterrestrial beings, for instance from some kind of engineering activity.

They used receivers widely spaced in the Soviet Union, working at wavelengths of 50, 30, and 16 centimetres. Starting in March 1970 they immediately began to detect coincident pulses at 50 centimetres wavelength during the day. To pin down the origin of these signals, stations were set up in Gorky, at Mount Karadag in the Crimea, Murmansk, and the Ussuri region of Siberia. These four stations covered 1,500 kilometres of longitude and 8,000 kilometres of latitude, so that random fluctuations caused by local interference would be distinguishable from broader phenomena.

From September to November 1970 the Soviet radio astronomers operated their receiving-stations for periods of two days on, followed by two days off. 'We observed phenomena which may tentatively be broken down into four groups,' says Troitsky.

First there were large pulses lasting between ten seconds and several minutes, and received coincidentally at all stations. These stood out clearly from the background noise. Another type of coincident phenomenon was described by Troitsky as being like 'a noise storm' on the receiver records. On top of these storms were periodic blips, which provide the third type of emission he described. And, fourthly, the Soviet radio astronomers noted long-lived changes in background emission, without any noticeable fluctuations.

Analysing these results, the group found that almost all the significant coincidences occurred in daylight. The existence of these widespread coincidences, and their number, suggested to Troitsky that a global phenomenon was involved, but he attributed it to effects in the atmosphere that would probably prove traceable to solar activity. 'Therefore', he concluded, 'the experiment did not detect any sporadic emissions reaching us from the Galaxy.'

More recently, however, reports have come from the Soviet Union that suggest a less cautious attitude towards their present results. In 1973 the Soviet news agency reported the discovery at Gorky of radio signals from space 'that they have never received before'. According to the report, the signals consisted of pulses that last for several minutes and are repeated several times a day. 'More than 30 Soviet scientists are now searching in earnest for radio signals from extraterrestrial civilizations,' the report added meaningfully.

This rather bald announcement threw western scientists into a mild turmoil, particularly as Professor Samuel Kaplan, director of Gorky Research Institute, made it plain that the possible extraterrestrial origin of these signals was under consideration.

The signals were received by two networks of ground stations, operated by Vsevolod Troitsky and Nikolai Kardashev. Troitsky's apparatus was installed at his four existing stations, while Kardashev had set up listening-posts in the mountains of the Caucasus, Kamchatka and the Pamirs.

Instead of the familiar dish-shaped antennae, Kardashev used a spiral antenna which was wound into a cone shape about a metre in height. Troitsky's apparatus was smaller still.

Kardashev was said in a news report to have determined that the signals came from within the Solar System, and were definitely artificial in origin. The signals picked up by Troitsky, which were rather different, have been attributed to an interaction between solar radiation and the Earth's magnetosphere.

But one explanation advanced for the emissions detected by Kardashev was that they are being sent out by an artificial satellite that has strayed into our Solar System, perhaps with the intention of making contact with us. This was reminiscent of a similar idea at that time being actively investigated in Britain; but this will be discussed in detail in the next chapter.

However, scientists in the West, after some thought, were more prepared to attribute the signals to some form of military space-probe activity—perhaps an attempt at Earth-bound 'intelligence', and quite likely (and wryly) a U.S. satellite spying on the Soviet Union.

Other star-searching programmes, little known outside the astronomical community, have since been undertaken at radio observatories in the United States.

In 1973 the South-African born radio astronomer Gerrit Verschuur reported * negative results in a listening programme on 21-centimetres wavelength at the National Radio Astronomy Observatory, Green Bank—the same location as the original Ozma project. Verschuur monitored ten stars selected on the basis of their nearness to the Sun and their likelihood of nurturing life-bearing planets—including, of course, the initial Ozma stars, Tau Ceti and Epsilon Eridani, which have now become standards in the signal-searchers' book.

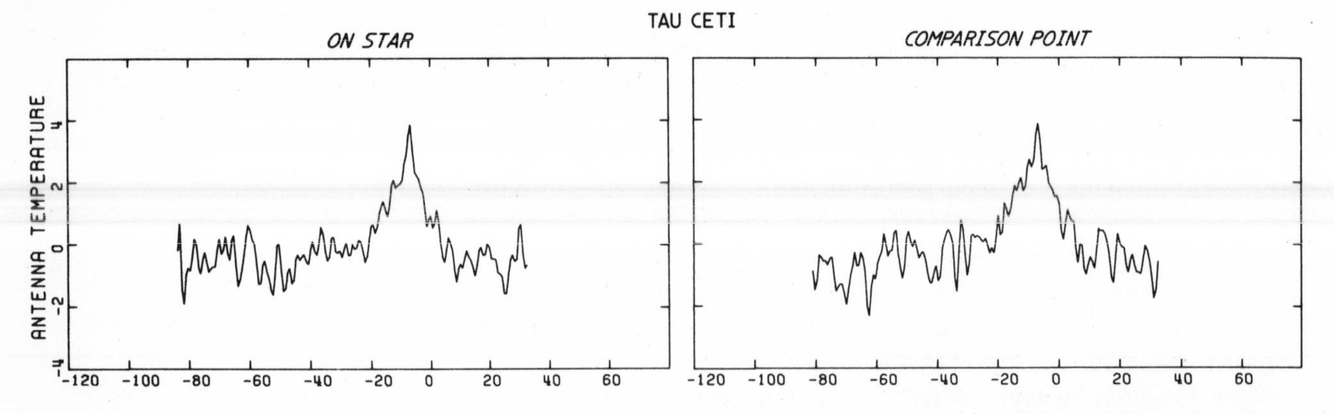

Radio noise from the direction of the star Tau Ceti and from a nearby area of blank sky shows no difference, thus proving that there is no signal being beamed towards us at or near 21-centimetre wavelength. (*Gerrit Verschuur*).

Verschuur had the advantage of being able to use the most sensitive 21-centimetre receiving equipment then available. The largest was the 300-foot-diameter dish, which Verschuur was using for a study of hydrogen clouds in the Galaxy. During his hydrogen-cloud observations he snatched quick glances at Tau Ceti, Epsilon Eridani and 61 Cygni in October 1971, observing each for a total of between one and two hours. Longer observing times, and the choice of other stars, were not possible because of the work programme of the telescope.

Verschuur estimated that the sensitivity of his receiver was a thousand times better than Troitsky's and that he would have detected a beacon transmission with an energy as low as 500 kilowatts.

Turning to the other major telescope now at Green Bank, the 140-foot radio dish, Verschuur re-examined those same three stars, plus 7 others including Barnard's star, during 1972. In these sessions, any transmissions with energies of a few megawatts should have been detected. Verschuur estimated that 5 minutes of his observing time equalled 4 days of observing with the Ozma equipment. Such is the increase in sensitivity of today's receivers.

The 140-foot radio telescope at Green Bank used in the search for possible signals sent by extraterrestrial beings. (*National Radio Astronomy Observatory*)

In trying to account for his negative result, Verschuur speculated that 21 centimetres might be the last place an intelligent civilization would be expected to broadcast on. They would probably keep it radio-quiet for their own astronomical observations, as we do on Earth.

But John Ball of Harvard had a different explanation. Taking a step

* *Icarus*, 19, 329.

into the realms of science fiction, he suggested that advanced civilizations, should they exist, have the same attitude towards conservation as we do, and that the Earth has been designated a wilderness area or sanctuary. Ball provocatively offered the idea that superintelligences are deliberately avoiding interaction with us because they have set aside the Earth as a zoo. 'They do not want to be found,' he asserted.

A more extensive search using the 300-foot and 140-foot Green Bank antennae was conducted by Benjamin Zuckerman and Patrick Palmer in 1972 and 1973, taking up a total of over 3 weeks' observing time. Frank Drake reported the results of their work to the annual meeting of the American Association for the Advancement of Science in March 1974. The radio astronomers made a systematic search of 600 nearby stars that seemed the likeliest centres of planetary systems like our own. But they failed to detect any radio emissions that would suggest the presence of technological civilizations. The stars that Zuckerman and Palmer monitored are thought to contain all the likely candidates for extraterrestrial life within 80 light years of the Sun. Frank Drake stated his belief that radio emissions from other civilizations 'certainly are now going through this room'. But, he emphasized, the problem was detecting them. He acknowledged the possibility that no one was actively trying to attract our interest, and that we may be forced to eavesdrop on stray signals from their domestic radio communications.

The 300-foot Green Bank radio telescope has been used to listen for signals from the stars. (*National Radio Astronomy Observatory*)

In a paper published in *Nature* in October 1973, Drake and Sagan discussed whether the 21-centimetre line of hydrogen really was uniquely ideal for interstellar contact. One other area of communication that had been suggested was the region of the spectrum bounded by the emission lines of hydrogen and the hydroxyl molecule (OH).

Communications engineer Bernard Oliver of the Hewlett-Packard Corporation has termed this 'the foreordained interstellar communication band' because H and OH go together to make water. Oliver asks: 'What more poetic place could there be for water-based life to seek its kind than the age-old meeting place for all species: the water hole?'

The so-called water hole is close to the region of minimum sky noise from all sources in the Universe, which adds to its attractiveness. Drake and Sagan point out in their *Nature* paper a specific frequency in the water hole at the centre of mass of H and OH, and they suggest that as the 'natural' communications frequency for water-based life. Life based on other substances would of course choose different frequencies, but these we could anticipate.

Another channel that Drake and Sagan find quite natural is specified by noise sources from the Universe. Minimum noise is found at a wavelength of around 5.4 millimetres—which would, however, require a radio telescope in space, as that wavelength is absorbed by the atmosphere.

The 150-foot Algonquin radio telescope in Canada is being used to listen for radio messages from the stars at the wavelength emitted by water molecules. (*National Research Council of Canada*).

Beginning in May 1974, two Canadian radio astronomers used the 150-foot Algonquin radio telescope to listen in, not at the water hole, but at the shorter 1.35 cm wavelength emitted by actual water molecules in space. Over a two-year period the astronomers, Paul Feldman and Alan Bridle, hope to look at up to 500 likely stars.

By the end of 1974 Feldman and Bridle had surveyed thirteen stars, including Barnard's star, Tau Ceti and Epsilon Eridani, and were preparing to cover one or two hundred more in late 1975.

At Arecibo in Puerto Rico is the 1,000-foot-wide radio telescope of America's National Astronomy and Ionosphere Center. This telescope, now operated by Cornell University, has recently been re-equipped to increase its sensitivity so that it is now capable of communicating with an identical radio dish anywhere in our Galaxy. This electrifying advance in technical capability has thrown wide open the doors of possible interstellar communication—and, as we shall see at the end of this chapter, a dramatic attempt to demonstrate this fact was not long in coming.

The Arecibo telescope is now being used by Frank Drake and his colleagues in a search for extraterrestrial signals. And a still larger radio telescope, 600 metres across, now being built in the Soviet Union has as one of its stated intentions, the search for interstellar beacons.

It is also an open secret among radio astronomers that a scan of the observing log at many radio observatories would reveal small snatches of time spent looking at individual stars, in hopes of catching an interstellar message.

These programmes go on despite the dispiriting results to date because, as we have seen, the total number of stars that we need to scan amounts to millions. Since only a few hundred stars have so far been monitored, the expected chance of success in the efforts to date is no more than one in ten thousand. So the radio astronomers clearly have a long way to go yet. And the most distant stars, where the chances of a signal are highest, have yet to be scanned.

Radio astronomer Robert Dixon of Ohio State University has voiced some ideas for a search strategy to find extraterrestrial radio beacons. * He believes that a civilization would place its beacon where the signal would have the greatest chance of being picked up by other radio astronomers. This, for him, means that a wavelength close to the 21-centimetre hydrogen line is the most likely choice.

In addition, says Dixon, the communicative civilization would want to transmit all the time in all directions, rather than hopefully beaming a signal at individual stars from time to time. Other astronomers have shown, quite rightly, that any other method of transmission leads to unrealistically long search times before success can be expected—that is in excess of a million years. As another ploy to optimize reception of the signal, Dixon imagines that the civilization would transmit continuously, not in on-off pulses. Information would be carried on the signal by the use of circular polarization, alternately left and right-handed.

What is our best strategy for finding such a beacon? Out to a distance of about 100 light years we could look at stars individually—there is only about one star this close to us per 25 square degrees of sky area. But at greater distances, out to 1000 light years, the number of stars increases to around 40 per square degree. To monitor that many in a reasonable time we should need to make broad, continuous sweeps.

I have suggested what might be a suitable strategy for interstellar signalling. Sending a beacon signal omni-directionally I believe to be impossible, because of the vast power needed in order to be heard over long distances. Neither do I believe that anyone would consider pointing an individual aerial at each of the likely hundred thousand or so target stars, because it is wasteful. Instead, I foresee that we might set up a cluster of partially-steerable aerials in each hemisphere to sweep over the plane of the Galaxy every few minutes. As most of the stars to which we wish to signal lie along the galactic plane, this is the most efficient way of broadcasting. To reach nearby stars which are clustered round us in all directions, we might set up weaker omnidirectional aerials in space, shielded from Earth to prevent radio noise at the wavelength we choose.

We could begin a limited version of this strategy today, covering a

* *Icarus*, October 1973, p. 187.

considerable percentage of the stars in the galactic plane. The signalling could be extended to wider areas as our technology improves over the next fifty years or so. If other civilizations use this strategy, which seems the most efficient yet proposed, then to stand the best chance of picking up their signal we should spend a few minutes listening to each area of the sky before moving on to the next. This could be the basis of an international listening project, with various countries allocated different areas of the sky, as in the Carte du Ciel project of optical astronomy.

Sagan believes that a 'brute force' method of listening to one star at a time is the best strategy for picking up a signal by deliberate effort. Such a programme should ideally be automated to increase its speed of operation, and be interspersed with conventional radio astronomy to relieve the scientists' frustration: 'It's very boring to go 30 years with no results,' Sagan comments in recognition of the enormous time factor involved.

Sums performed by radio engineers have produced the remarkable conclusion that, at high radio frequencies, the Earth is a far brighter object than our Sun. The main culprit is our television transmissions. As a result, some astronomers have enthused about the possibilities of eavesdropping on the domestic communications of other civilizations. This would, indeed, be the only way to detect extraterrestrial life if everyone listened for communications without sending any.

Frank Drake has shown that, even with the most advanced statistical techniques of examining a signal, there would be a reduction of ten thousand times in sensitivity in this eavesdrop method, as compared with direct reception of a beacon signal. We could still pick up radio noise like our own out to a distance of perhaps 100 light years if we were prepared to build a sufficiently big (and expensive) collecting area.

The full Cyclops array, here visualized by an artist, would cover 20 square kilometres or more. Each dish would be of 100 metres diameter, and the control centre would be located at the centre of the array. (NASA)

Carl Sagan pours cold water on this optimism by an assessment of our telecommunications capacity. He points out that, throughout its 4,500-million-year history, the Earth has been radio quiet until the invention and exploitation of radio communication in the past two human generations.

'But now we're moving to tight-beam transmissions, synchronous satellites, and cable television,' he says. 'It's clear that soon the radio brightness of the planet Earth will decline.

'I think there's just a 100-year-long spike in radio emissions before the planet becomes radio quiet again. So the chance of success in eavesdropping for a given inhabited planet is 100 divided by 4,500 million.

'Which', he adds dryly, 'is negligible.'

Sagan is firmly convinced from his statistical calculations reported in Chapter One that there are not likely to be civilizations close to us in space that are also close to us in time.

One possible super-receiver for picking up interstellar transmissions was discussed in 1971 at a design-study meeting at NASA's Ames Research Center attended by communications engineers and astronomers. They detailed their speculations and findings in a 243-page report entitled *Project Cyclops*. The Cyclops team concluded that it is technically possible to build far larger radio-collecting arrays than we currently have available. Though very big individual dishes are not ruled out, in space or on the ground, we can programme a whole field of smaller dishes to act in unison, giving a combined collecting-area equivalent to one massive dish with a width of several kilometres.

Large-size collecting arrays such as this are necessary in order to establish the initial contact. Assuming a power of 1,000 megawatts for the beacon transmitter, the Cyclops team calculated that their proposed super-array could make contact over a distance of up to 1,000 light years. Clearly, more powerful beacons than this would be detectable over greater distances. And weaker signals—perhaps leakage from domestic communications—could be picked up from nearer stars.

Unit-building the Cyclops system has the advantage that its overall size can be docked or extended at will. Its search programme will begin as soon as the first few antennae are operating, and will be continued deeper and deeper into space as its collecting area is added to. Naturally, should the Cyclops search prove successful at any intermediate stage of its construction, the rest of the array need not be built.

Bernard Oliver, one of the directors of the Cyclops study, has visualized the final system as 'an orchard of antennae covering 20 square kilometres or more, all feeding a single data processing system and all under automatic computer control'. The individual dishes would be 100 metres across (which is still very big in today's terms), and there would be at least 1,000 of them. 'The total construction time would be 10 to 20 years,' says Oliver.

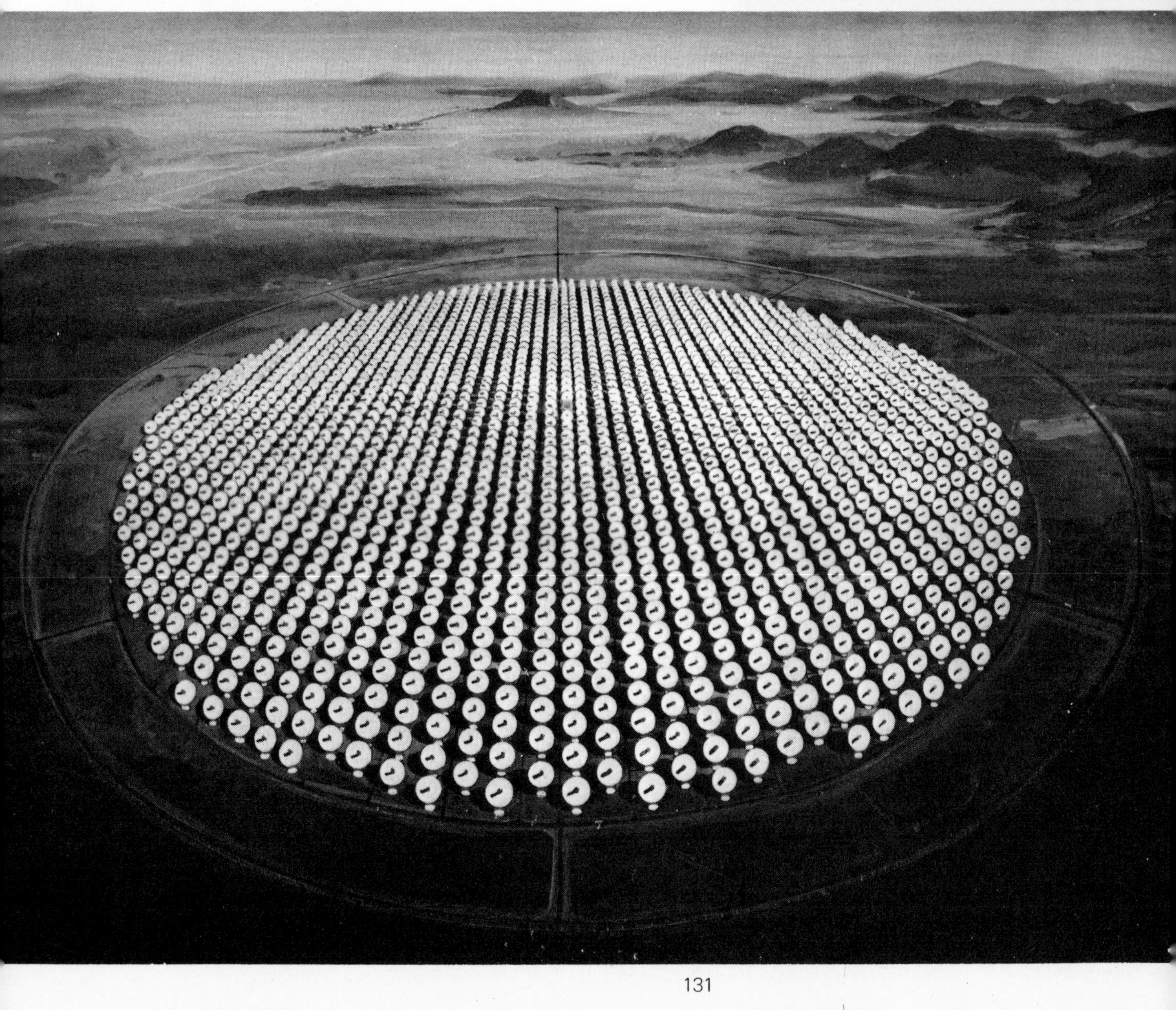

The only limiting factors are our engineering ability—and the cost. A broad estimate is that Cyclops would absorb around $600 million during each year of its construction, giving it a total bill probably in excess of the Apollo Moon programme.

Oliver calculates that the Cyclops system has several million times the sensitivity of the Ozma search, and can scan the spectrum 200,000 times faster. According to the Cyclops report, any inhabitants of Tau Ceti or Epsilon Eridani would have had to be emitting around 10 million megawatts to wiggle the pens of Ozma's recorders. The Cyclops system, by comparison, could detect 2.5 megawatts.

Oliver is quick to point out that the massive cost of Cyclops would not be spent purely for an all-or-nothing attempt at finding our hypothetical neighbours. The instrument could, and would, be usable as an ordinary radio telescope of unprecedented capability, both to study faint emissions from the Universe and to provide radar mapping of the planets with far greater detail and range than is currently possible.

The enormous collecting-area of Cyclops would also provide vastly improved ranges and rates of transmission with interplanetary—and perhaps even interstellar—probes. One other particularly topical suggestion is that the Cyclops reflectors could be used to provide solar power during the day.

One problem highlighted by the Cyclops study team is the lack of a comprehensive catalogue of all the likely target stars within 1,000 light years of the Sun. In short, we don't yet know the best places to look for life outside of the few hundred nearest stars which optical astronomers have studied in detail. So the Cyclops team recommends an optical search for all the stars like the Sun out to the adopted maximum range of 1,000 light years, to prevent valuable radio-telescope time being wasted by listening to stars with no chance of having life-bearing planetary systems.

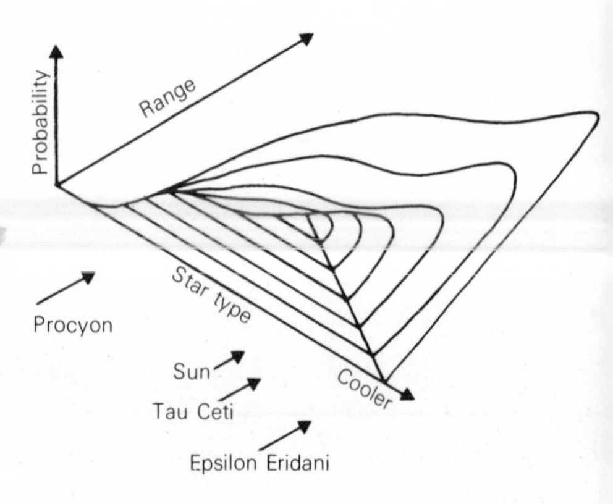

The Cyclops team plotted this graph to show their suggested search strategy for messages from different types of star. The probability of success is highest for Sun-type stars, so these would be searched out to greater distances. For stars less likely to harbour life, only those nearby would be bothered with.

The Cyclops programme would begin as soon as the first antennae were installed by concentrating on the nearby stars. Then, as the array was increased and the optical astronomers improved their catalogues of interesting target stars, the search would be pushed out to ever-greater distances. Part-way through the search, suggests the Cyclops report, the instrument might be used to send a beacon signal for a year or more. The targets would be the nearest stars, which we could then monitor for any response after an appropriate lapse.

The final Cyclops array would satisfy one of the recommendations of the Byurakan conference on communication with extraterrestrial intelligence (CETI). The delegates to the conference concluded that a variety of instruments working at various wavelengths would be desirable in the search for evidence of extraterrestrial life.

One of the proposed systems was a radio telescope with a collecting-area of a square kilometre or more. Other instruments they proposed would work at shorter wavelengths, and be correspondingly smaller, but all delving into the short radio or infra-red bands to search for evidence of extraterrestrial activity.

One of the most exciting developments in physics during the past twenty-five years has been the invention of amplifiers capable of emitting intense beams of radiation at certain specified wavelengths. In the short-radio region of the spectrum, where they were first made, such devices are termed masers. It is the maser that has made possible the sensitivity of today's radio telescopes and satellite-communication aerials.

The 200-inch optical telescope on Mount Palomar. (U.S.I.S)

Charles Townes, the American physicist who produced the first maser, was one of those who foresaw that the same principle could be applied to light-beams, creating what is now termed a laser. By the early 1960s, various types of laser were in action throughout the world. In 1961 Townes put his name to a suggestion for interstellar communication by laser. He suggested we could use, for instance, the mirror of a giant optical telescope, such as that of the Mount Palomar 200-inch, to focus and reflect the intense beam of a laser towards the stars — effectively reversing the telescope's usual function.

Alternatively, we might bunch lasers together into an array that shot a combined light-beam directly into space.

Though this would not have nearly so much penetration as the telescopic idea, it would be easier to mount above the atmosphere — and a transmitting post in space was, Townes said, essential to prevent the smearing and filtering effect of the Earth's atmosphere.

How could we make a laser-beam visible above the swamping brilliance of our Sun? The suggested answer relies on those dark lines in the spectrum by which astronomers analyse the composition of stars. Particularly prominent dark lines are caused in the Sun's spectrum by calcium. If we could arrange a laser-beam whose emission was centred at precisely such a wavelength, we would produce an effect that astronomers of another civilization could recognize as being clearly artificial.

Radio astronomers have not been impressed by the idea of signalling at optical frequencies. The Cyclops team made a detailed comparison between signalling by laser and signalling by microwave radio transmissions. 'The verdict is heavily in favour of microwaves', reports Bernard Oliver.

Simple advantages of radio waves are that they can be detected at all times of day and in all weathers. More technically, the power needed for long-distance communication is much greater in lasers than for microwaves. Large collecting-areas for light-waves are much more difficult and expensive to build than large radio arrays. And, if we want to transmit a beacon-signal in all directions to optimize the chances of being detected, paradoxically the main advantage of lasers — their narrow beams — is actually a drawback.

The laser supporters are not silenced by this opposition. At the Byurakan conference, Charles Townes emphasized the rapid development of power outputs in lasers. He presented arguments to show that an infra-red laser is a reasonable choice for communication, at least with the nearest stars. He thought that radio might not be the only logical choice for interstellar communication. 'It is important to recognize that we may not have thought imaginatively or with enough conviction about some of the other methods,' he said. There might be situations, such as on high-gravity planets with strong storms, where the environment precluded the building of large antennae.

In view of the fact that finding life on nearby stars would allow exchange of messages in a human lifetime, Townes is in favour of a constant watch on stars within about 10 light years.

One man who has thought about the use of lasers for interstellar signalling is British communications engineer Tony Lawton. In 1971 he pointed out that the most efficient lasers work in the infra-red region of the spectrum, and proposed focusing the beam of such a laser on to the nearer stars via giant mirrors hundreds of feet wide. A mirror of this size could not be made from conventional materials, but might instead be constructed in the same way as lightweight mirrors for space probes: that is, using reinforced resin, which is spun in a liquid state until it reaches the right curvature, and then cured so that it solidifies into a mirror shape. It is then coated with a reflecting layer. Surfaces sufficiently accurate for infra-red waves can be made in this way.

Probably such a telescope would need to be placed on the Moon, but Lawton calculates that a signal from such a set-up could be detected by a similar-sized telescope up to 500 light years away.

Lawton had already proposed building giant infra-red mirrors to search for planets round nearby stars. If, on such planets, an intelligent civilization were beaming an infra-red laser our way, the signal he said, 'would come roaring into the sensitive detector system and give the effect of an infra-red double star where visually only a single star exists.' A methodical planet-hunting survey as proposed by Lawton would therefore probably detect any such infra-red signal, especially if it were slotted into a particular wavelength where strong radiation from the parent star was not expected.

He put it that the economics of such a system actually argued in favour of its usage over radio signalling. A one-megawatt laser installation, said Lawton, would probably cost less than 5 per cent of its radio counterpart. And he urged that an infra-red survey should be undertaken on the regions round the nearest stars.

One of the possible sources of an infra-red signal could be a 'stray' planet that had been spun off from its parent star during the period of planetary formation, and had continued its isolated course ever since.

A planet such as Jupiter with its own internal energy source might give rise to life even without the light from a parent star. For the inhabitants of such a planet, visible light would be an alien wavelength; they would communicate at infra-red and radio wavelengths. In 1974 Tony Lawton argued in a paper published in the British magazine *Spaceflight* that the possibility of a pulsating infra-red or radio signal, coming from the direction of no visible star, should not be overlooked in sky surveys.

One other rather bizarre mode suggested for drawing the attention of other civilizations is the seeding of a star's atmosphere with quantities of some rare element that would produce unexpected lines in the star's spectrum. The practical difficulty of doing this, as well as the limited range at which it would be detectable, militate against it.

Assuming that we are able to detect some sort of beacon-signal from the stars, what would we expect it to say?

A series of numbers, transmitted in the Morse-like code of binary symbols used by computers, would certainly distinguish the signal from any possible source of random noise. Such a suggestion has been airily dismissed on the grounds that it would simply prove that the senders could count, which we might in any case assume. Stanford University's Ronald Bracewell felt instead that a continuous signal would be most likely for pre-contact signalling.

From a stable beacon on a planet orbiting a star, he believed, we could deduce the planet's day and year, the size of the orbit and of the planet itself, and perhaps even the beacon's latitude. 'As a first harvest during the interval before a signal could be acknowledged, this would be much more exciting news than a string of prime numbers,' he wrote in the *Proceedings of the Institute of Radio Engineers* in 1962.

Communications engineer J.A. Webb of the Lockheed Georgia Company foresaw that similar opportunities would be opened up by eavesdropping on microwave signals used for anti-missile radar, interplanetary surveillance and space communications of other civilizations. 'It provides essentially the only hope that within the lifetimes of some of us it may be possible to look into the lives of alien intelligent beings,' he said at a symposium of radio engineers in 1961.

But most of the thinkers on extraterrestrial communication expect an interstellar signal to carry some form of intelligent message. 'It seems quite likely that a television picture would be the logical way to open an interstellar conversation,' says astrophysicist Alastair Cameron.

A string of binary pulses that, when displayed as a pictorial array, shows the crude outline of a human being has now become a cliché in the literature on interstellar communication. Yet such messages can easily be conveyed in a few digits.

In simple pictures we could learn about the nature of the senders' civilization: their solar system, their physical shape and even the biochemistry of their bodies. The stitches of the Bayeux tapestry and the tiles in a mosaic form the individual bits of a digital picture. So pictures formed from pulses are indeed a powerful yet simple way of transmitting complex ideas.

Mathematicians have gone to considerable lengths to invent 'logical' languages, like the languages of computers, which would make the transmission of more subtle information possible. In effect, we are trying to build a code that is as easy as possible for anyone to crack; this has been termed anticryptography. Unless they are being deliberately obscure—perhaps to prevent naïve people such as ourselves from joining in the conversation—we may assume that other civilizations trying to attract attention would do the same.

Participants at the Byurakan symposium debated the problem of message deciphering in a special session. Soviet radio astronomer B.I. Panovkin disagreed with Philip Morrison about the ease with which two civilizations would be able to decrypt each others' messages. Panovkin argued that for one civilization to understand the symbols of another, the two worlds would need to be very similar in historical background. 'Semantic communication is very limited,' he declared. Morrison proposed that early in the communication would come a kind of cosmic Rosetta Stone that, once decoded, would allow us to understand the whole message. Most of the conferees were willing to accept this point of view, but Panovkin disagreed that mathematics is universal. 'There may be different axiomatic bases of mathematics,' he ventured.

Despite this piece of scientific agnosticism, Soviet linguist B.V. Sukhotin has been preparing computer programs to test signals from space for possible messages. The programs are designed to reveal whether or not the signals are artificial in nature and if so, to attempt to decode them.

Philip Morrison pointed out that information about all of Greek culture could be transmitted in about 100 seconds. This is far too rapid for human reading, and so would be stored on tape for later decipherment. He gave his view of how we would absorb the message: 'Not as one reads the newspaper but as one works out a rich, difficult textbook of an advanced subject full of diagrams, hints, and examples.'

Once a machine had a translation program stored within it, the sending and receiving of interstellar messages would be technologically straightforward.

In fact, the Byurakan conferees were remarkably conservative in their assessment of the future role to be played in societies by intelligent machines. Discussing these aspects in their book *Intelligence in the Universe*, Roger MacGowan and Frederick Ordway proposed that most of the advanced technological societies with which we might communicate will in fact be directed by a mechanical super-brain. 'Moreover, totally independent intelligent automata, created by extinct or abandoned biological societies, must also be prevalent,' they added. For them, it was an 'inescapable conclusion' that the communications we would one day receive would come from one of these two sources, rather than a purely biological society.

But, so far, no one seems to have modified the Drake formula to take such cybernetic civilizations into account. Improved recognition of our technological potential will inevitably colour the CETI debates in the future.

At Byurakan, Marvin Minsky conjured up the spectre of Fred Hoyle's famous novel *A for Andromeda* in which a less-than-benevolent society transmits the plans for a super-computer containing its own program. This could be a very insidious way of sending a missionary (or a dictator) from one civilization to another.

From work of his own and of colleagues, Minsky is now convinced that it is possible to make a computer program that can interact with someone else's language. 'It appears that in 1970 we crossed a threshold of being able to deal with semantics in computers,' he says.

The moral aspects of interstellar contact came uppermost in the minds of the participants at the end of the Byurakan conference. 'We need from the beginning to accept responsibility for the effect of what we may be planning,' said Morrison. He argued that the slow decoding and interpretation of a complex message from space, which would involve many scholars and many books of commentary, would act as a kind of buffer to the cultural shock that many have feared from interstellar communication. The very lack of physical communication would prevent any fear of our military or commercial dominance by advanced civilizations.

Morrison believed that the message would contain little information on science and mathematics; we would be able to solve those problems more easily for ourselves, rather than relying on extraterrestrial messages. Instead, he thought, the message would contain a rich cultural tapestry of art and history, giving a glimpse of a distant future that we cannot imagine: 'It will produce a view of the Universe less alienating than the one now characteristic of the general population of industrial societies.

'Such a signal will never come to dominate all of human experience but will nevertheless be a very rich contribution.'

Historian W.H. McNeill expressed the view that mutual intelligibility between the civilizations of two planets would be 'very strained' as a result of differences in language, experience and physical make-up. And he warned, 'Contact between men has shown that those who had the power used it.'

Sebastian von Hoerner drew the parallel between the meeting of a modern Earth culture and a Stone Age culture; in time, the Stone Age culture is always swamped. The same, he believed, would happen as a result of our meeting with the ideas of a superior civilization: 'Our period of culture would be finished and we would merge into a larger interstellar culture.' (Whether or not this would be a desirable development is subjective.)

Others echoed the view that knowledge of other civilizations in space would be reassuring to Earth's inhabitants. Says Czech scientist Rudolf Pesek, 'My personal opinion is that CETI can help us to solve our terrestrial problems and increase the lifetime of our civilization.'

One advantage of extraterrestrial contact may be to mute the parochial attitudes among the backwoodsmen of contemporary society. Nationalism is already outdated. Instead of internationalism, the new outlook would be interstellarism.

Su-Shu Huang has suggested that to get any sensible message from another civilization, we may first have to reply to their probing signal: 'They cannot waste their time to have a long talk with mute people.'

Reception of a signal necessarily raises the problem of whether we should answer it. Though Soviet astrophysicist V.L. Ginzburg has claimed that any danger is 'extraordinarily improbable', we saw from the views of Zdenek Kopal in the Introduction that not all scientists agree.

The Project Cyclops report summarized the possible hazards and concluded: 'Before we make such a response or decide to radiate a long-range beacon, we feel the question of the potential risks should be debated and resolved at a national or international level.'

Tony Lawton has pondered the morality of talking to the stars. He says: 'The CETI experiments to date in my opinion have been designed almost with the will to fail—the people have not stated specifically in their programmes what they will then do if they succeed.

'I for one would like to believe that, if there is success in these experiments, the world is informed as soon as the people concerned are sure that they have received a CETI signal. I am rather concerned that one nation may attempt unilateral CETI.'

He regards the first intelligent signals from elsewhere as presenting a test for man that is much stiffer than mere technical knowledge and ability: 'He may have to give the first truly honest and unbiased answers in his history.'

Yet, of course, our radio noise is already circulating in the whispering-gallery of the cosmos. And our first deliberate messages are already on their way to the stars. First to be sent were the identical plaques attached to the Pioneer 10 and 11 spacecraft, which are even now silently gliding among the planets on trajectories that will ultimately eject them from the Solar System for ever.

The plaque, designed by Carl Sagan and Frank Drake, with the assistance of Sagan's wife artist Linda Sagan, has easy and difficult parts for another civilization to decipher.

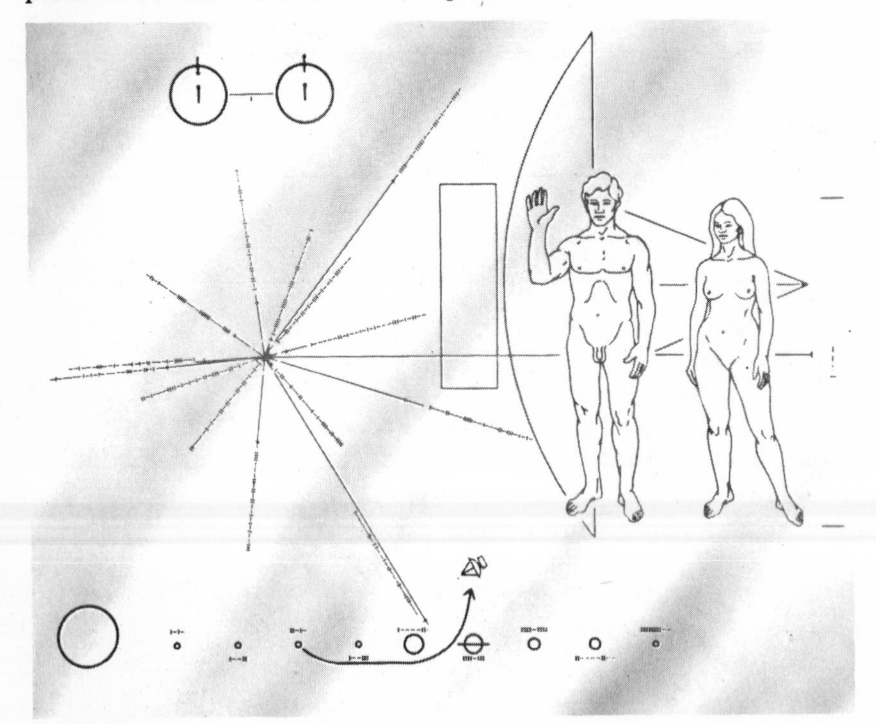

The engraved plaque on the Pioneer 10 and 11 space probes. From the Ian Ridpath/David Hardy filmstrip *Exploring the Planets*. (*Hulton Educational Publications*).

The easy parts are the representation at top left of hydrogen atoms, which of course emit the standard 21-centimetre radiation, and the diagram left of centre that shows the radial distribution of pulsars as seen from the Sun. The tick marks denote the period of each of the pulsars.

Another civilization would be able to reconstruct the approximate position of our Sun from the directions in which we see the pulsars. Since pulsars run down with time, the rough date of launch of the spacecraft could also be deduced by comparing the periods as marked on the plaque with the periods as observed at the time of intercepting the probe.

At the bottom of the plaque is a schematic representation of our Solar System, with the outward path of Pioneer marked.

A rough outline of the probe is given at the right, which will of course be recognizable to anyone who intercepts it. But there is also a difficult part of the message.

These are two figures, whose size can be gauged by comparison with the size of the probe's outline. But what do the figures represent?

We can recognize them easily, but creatures of a totally different cast may well find this the most baffling part of the plaque. The man has his hand raised in what is intended to be a symbolic gesture of friendship. But it would be foolish to imagine that every civilization in the Universe would see it this way.

By the designers' own admission, the chances of the probes ever being intercepted are small. They are aimed at no star; they are unlikely ever to stray into another planetary system. We can only hope that a civilization in space may have sufficiently sensitive surveillance techniques to detect and collect such an interstellar straggler.

The Pioneers will drift dead and dark between the stars long after the Sun has swollen and the Earth is dead. Man's longest-lasting artefacts will have drifted across the Galaxy by that time.

The message of the plaques is unlikely to be a dangerous give-away, because compared with the sunbeam speed of a radio wave the Pioneers are only snailing into space.

The gold-plated Pioneer plaques are like bottled messages bobbing on the interstellar ocean. They may never be seen by anyone else.

But they have been closely analysed on Earth. Echoing the theme of this book's Introduction, Sagan says, 'The transmittal of the Pioneer 10 message encourages us to consider ourselves in cosmic perspective.' And he adds: 'The greatest significance of the Pioneer 10 plaque is not as a message to out there; it is a message to back here.'

Most remarkable of all, though, was a three-minute message transmitted as part of the re-dedication ceremony of the Arecibo radio telescope on November 16th 1974. As mentioned previously, modifications to the Arecibo telescope mean that it can now, in theory, communicate with a similar instrument anywhere in our Galaxy. A three-minute radio burst can hardly be considered a major attempt at interstellar communication; as with the Pioneer plaques, its significance may be greatest here on Earth, awakening people to the possibility of impending interstellar contact and our ability to reply.

The Arecibo radio telescope, sporting its newly resurfaced dish, lies in a natural hollow in the mountains of Puerto Rico. (*National Astronomy and Ionosphere Center*)

```
0 0 0 0 0 0 1 0 1 0 1 0 1 0 0 0 0 0 0 0 0 0 0 0 0 1 0 1 0 0 0 0 0 1 0 1 0
0 0 0 0 0 0 1 0 0 1 0 0 0 1 0 0 0 1 0 0 0 1 0 0 1 0 1 1 0 0 1 0 1 0 1 0 1
0 1 0 1 0 1 0 1 0 1 0 0 1 0 0 1 0 0 0 0 0 0 0 0 0 0 0 0 0 0 0 0 0 0 0 0 0
0 0 0 0 0 0 0 0 0 0 0 0 0 0 0 0 1 1 0 0 0 0 0 0 0 0 0 0 0 0 0 0 0 0 0 0 0
1 1 0 1 0 0 0 0 0 0 0 0 0 0 0 0 0 0 0 0 0 0 0 1 1 0 1 0 0 0 0 0 0 0 0 0 0
0 0 0 0 0 0 0 0 1 0 1 0 1 0 0 0 0 0 0 0 0 0 0 0 0 0 0 0 0 0 0 1 1 1 1 1 0
0 0 0 0 0 0 0 0 0 0 0 0 0 0 0 0 0 0 0 0 0 0 0 0 0 0 0 0 0 0 0 1 1 0 0 0 0
1 1 1 0 0 0 1 1 0 0 0 0 1 1 0 0 0 1 0 0 0 0 0 0 0 0 0 0 0 0 0 1 1 0 0 1 0
0 0 0 1 1 0 1 0 0 0 1 1 0 0 0 1 1 0 0 0 0 1 1 0 1 0 1 1 1 1 1 0 1 1 1 1 1
0 1 1 1 1 1 0 1 1 1 1 1 0 0 0 0 0 0 0 0 0 0 0 0 0 0 0 0 0 0 0 0 0 0 0 0 0
0 1 0 0 0 0 0 0 0 0 0 0 0 0 0 0 0 0 0 1 0 0 0 0 0 0 0 0 0 0 0 0 0 0 0 0 0
0 0 0 0 0 0 0 0 0 0 0 1 0 0 0 0 0 0 0 0 0 0 0 0 0 0 0 0 0 1 1 1 1 1 1 0 0
0 0 0 0 0 0 0 0 0 0 0 1 1 1 1 1 0 0 0 0 0 0 0 0 0 0 0 0 0 0 0 0 0 0 0 0 0
0 0 1 1 0 0 0 0 1 1 0 0 0 0 1 1 1 0 0 0 1 1 0 0 0 1 0 0 0 0 0 0 0 1 0 0 0
0 0 0 0 0 0 1 0 0 0 0 1 1 0 1 0 0 0 0 1 1 0 0 0 1 1 1 0 0 1 1 0 1 0 1 1 1
1 1 0 1 1 1 1 1 0 1 1 1 1 1 0 1 1 1 1 1 0 0 0 0 0 0 0 0 0 0 0 0 0 0 0 0 0
0 0 0 0 0 0 0 0 0 1 0 0 0 0 0 0 1 1 0 0 0 0 0 0 0 0 0 1 0 0 0 0 0 0 0 0 0
0 0 1 1 0 0 0 0 0 0 0 0 0 0 0 0 0 0 0 1 0 0 0 0 0 1 1 0 0 0 0 0 0 0 0 0 0
1 1 1 1 1 1 0 0 0 0 0 1 1 0 0 0 0 0 0 1 1 1 1 1 0 0 0 0 0 0 0 0 0 0 1 1 0
0 0 0 0 0 0 0 0 0 0 0 0 1 0 0 0 0 0 0 0 0 1 0 0 0 0 0 0 0 0 1 0 0 0 0 0 1
0 0 0 0 0 0 1 1 0 0 0 0 0 0 0 1 0 0 0 0 0 0 0 1 1 0 0 0 0 1 1 0 0 0 0 0 0
1 0 0 0 0 0 0 0 0 0 0 1 1 0 0 0 1 0 0 0 0 1 1 0 0 0 0 0 0 0 0 0 0 0 0 0 0
0 1 1 0 0 1 1 0 0 0 0 0 0 0 0 0 0 0 0 0 1 1 0 0 0 1 0 0 0 0 1 1 0 0 0 0 0
0 0 0 0 1 1 0 0 0 0 1 1 0 0 0 0 0 0 1 0 0 0 0 0 0 0 1 0 0 0 0 0 0 1 0 0 0
0 0 0 0 0 1 0 0 0 0 0 1 0 0 0 0 0 0 0 1 1 0 0 0 0 0 0 0 0 1 0 0 0 1 0 0 0
0 0 0 0 0 1 1 0 0 0 0 0 0 0 0 1 0 0 0 1 0 0 0 0 0 0 0 0 0 1 0 0 0 0 0 0 0
1 0 0 0 0 0 1 0 0 0 0 0 0 0 1 0 0 0 0 0 0 0 1 0 0 0 0 0 0 0 1 0 0 0 0 0 0
0 0 0 0 0 0 1 1 0 0 0 0 0 0 0 0 0 1 1 0 0 0 0 0 0 0 0 1 1 0 0 0 0 0 0 0 0
0 1 0 0 0 1 1 1 0 1 0 1 1 0 0 0 0 0 0 0 0 0 0 0 1 0 0 0 0 0 0 0 1 0 0 0 0
0 0 0 0 0 0 0 0 0 0 1 0 0 0 0 0 1 1 1 1 1 0 0 0 0 0 0 0 0 0 0 0 0 1 0 0 0
0 1 0 1 1 1 0 1 0 0 1 0 1 1 0 1 1 0 0 0 0 0 0 1 0 0 1 1 1 0 0 1 0 0 1 1 1
1 1 1 1 0 1 1 1 0 0 0 0 1 1 1 0 0 0 0 0 1 1 0 1 1 1 0 0 0 0 0 0 0 0 0 1 0
1 0 0 0 0 0 1 1 1 0 1 1 0 0 1 0 0 0 0 0 0 1 0 1 0 0 0 0 0 1 1 1 1 1 1 0 0
1 0 0 0 0 0 0 1 0 1 0 0 0 0 0 1 1 0 0 0 0 0 0 1 0 0 0 0 0 1 1 0 1 1 0 0 0
0 0 0 0 0 0 0 0 0 0 0 0 0 0 0 0 0 0 0 0 0 0 0 0 0 0 0 0 0 0 0 0 1 1 1 0 0
0 0 0 1 0 0 0 0 0 0 0 0 0 0 0 0 0 0 1 1 1 0 1 0 1 0 0 0 1 0 1 0 1 0 1 0 1
0 1 0 0 1 1 1 0 0 0 0 0 0 0 0 0 1 0 1 0 1 0 1 0 0 0 0 0 0 0 0 0 0 0 0 0 0
0 0 1 0 1 0 0 0 0 0 0 0 0 0 0 0 0 0 0 1 1 1 1 1 0 0 0 0 0 0 0 0 0 0 0 0 0
0 0 0 1 1 1 1 1 1 1 1 1 0 0 0 0 0 0 0 0 0 0 0 0 1 1 1 0 0 0 0 0 0 0 1 1 1
0 0 0 0 0 0 0 0 0 1 1 0 0 0 0 0 0 0 0 0 0 0 1 1 0 0 0 0 0 0 0 1 1 0 1 0 0
0 0 0 0 0 0 0 1 0 1 1 0 0 0 0 0 1 1 0 0 1 1 0 0 0 0 0 0 0 1 1 0 0 1 1 0 0
0 0 1 0 0 0 1 0 1 0 0 0 0 0 1 0 1 0 0 0 1 0 0 0 0 1 0 0 0 1 0 0 1 0 0 0 1
0 0 1 0 0 0 1 0 0 0 0 0 0 0 0 1 0 0 0 1 0 1 0 0 0 1 0 0 0 0 0 0 0 0 0 0 0
0 1 0 0 0 0 1 0 0 0 0 1 0 0 0 0 0 0 0 0 0 0 0 0 1 0 0 0 0 0 0 0 0 0 1 0 0
0 0 0 0 0 0 0 0 0 0 0 0 1 0 0 1 0 1 0 0 0 0 0 0 0 0 0 0 0 1 1 1 1 0 0 1 1
1 1 1 0 1 0 0 1 1 1 1 0 0 0
```

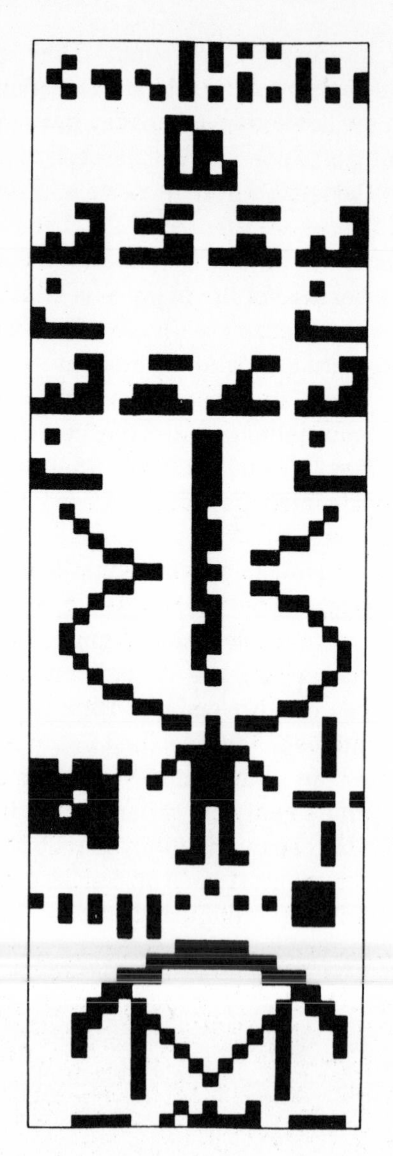

The 1,679 digits of the Arecibo message, transmitted on November 16th 1974. The 1's and 0's, transmitted as on-off pulses like in Morse code, can be rearranged to give a pictogram message that describes human biochemistry and other information about our species (see text).

The Arecibo message consisted of 1,679 binary pulses, represented in the diagram as zeroes and ones, transmitted at a rate of ten a second towards a globular cluster of stars, in the constellation Hercules, named M13. The point of choosing such an object as a target is that the radio telescope's beam can cover all 300,000 stars of the cluster in one go. This is clearly a very economical arrangement, and it is surprising that globular clusters have received no previous attention from CETI researchers.

The M13 globular star cluster is about 24,000 light-years away, so there is no chance of a rapid reply. Carl Sagan estimates that there is a one-in-two chance of someone being there to receive it.

The key to the message is that it breaks down into a grid, 23 by 73 characters, in which a picture can be built up from the on-off pulses of the signal. This is similar in style to test-messages that radio astronomers have swapped among themselves to study the method's potential. As shown in the diagram here, the pictogram starts from the top right, and describes in binary form the numbers one to ten. This is intended as a lesson to the recipients, which establishes the language of the following message. Below this top row is a group displaying the atomic numbers of hydrogen, carbon, nitrogen, oxygen and phosphorus, the chemicals of life.

Next, the message uses this information to describe the molecular components of D.N.A., and on lines 32 through 46 actually depicts D.N.A.'s famous double-helix structure. The central core represents

Is there anybody there? The 300,000 stars of the M 13 globular cluster will receive the Arecibo message in 24,000 years' time. (*Hale Observatories*)

the number 4,000 million, which is roughly the number of characters in the genetic code.

Thus we have described the chemical basis of terrestrial life. Next comes a cryptic depiction of a human, with an indication of his height to the right (14 wavelengths of the transmission) and on the left the approximate number of the human population (4,000 million again, but written in a different way).

On the next line is a sketch of the Solar System, with below it a representation of the Arecibo telescope itself, pointing downwards to a number that roughly describes its 1,000-foot diameter.

This interstellar IQ test, which is the brainchild of Frank Drake and his colleagues, was arrived at after exhaustive discussions and, as Drake puts it, being 'market tested' to see how easy it was to decipher. For a layman it is baffling, even when described. A number of scientists were able to decipher most of it, but they were already familiar with the working of those minds that invented it. This recalls the point of B.I. Panovkin, who argued at Byurakan about the problem of communication between different galactic cultures.

The message was transmitted at a wavelength of 12.6 centimetres (frequency 2380 MHz)—not one of the standard lines suggested for interstellar communication, but instead a wavelength being used at Arecibo for radar observations of the planets. This, again, undermines its value as a serious attempt at interstellar signalling. The signal frequency was modulated to correct for the motion of the Earth in its orbit. However, as we have seen, other astronomers have argued that an uncorrected signal would reveal interesting extra information about the transmitting planet.

The Arecibo message was transmitted with no prior announcement to the scientific community. In view of the reservations about CETI reviewed in this chapter, and the Byurakan agreement that such things were best done 'by representatives of the whole of mankind', there was a predictably critical response. Answering my questions for an article in *Nature*, Drake said that he did not consider the message a sufficiently major event to require international cooperation. But Tony Lawton, who is a member of the CETI standing committee of the International Academy of Astronautics, thought that it had established 'a very nasty precedent', and was frankly surprised that it had been done without prior international discussion and approval—particularly since Drake was a leading signatory of the Byurakan resolutions. Lawton says: 'Now there is nothing to stop any nation from transmitting any signal, anywhere, to any spot it chooses. A little warning, a little consultation, would have been better.'

Drake would not commit himself on the chances of the message being received and understood. But there were, he said, plans to retransmit the signal when telescope time permitted.

8

TRAVEL TO THE STARS

We already have the ability to send space probes to the stars. But even the Pioneers on their high-speed flights would take eighty thousand years to reach the nearest. The problem, clearly, is one of time. How can we reach the stars in a reasonable fraction of human-life span?

The simple answer is to go faster; but that needs more power—vastly more than anything we have available at present. Although spacecraft have been put on interstellar trajectories within two decades of the launching of the first satellite, reaching the stars *quickly* may take all our ingenuity for the next two centuries.

The launching of Skylab showed that we have the ability to orbit a structure the size—if not the shape—of a three-bedroomed house. It would be possible with little more effort to put together a roomy space station and push it out towards the stars.

Such a hotel of the skies would be like the space ark first envisaged by J.D. Bernal, in which generations of space travellers live and die knowing no other home, until their descendants finally step out at the destination to found a new colony.

But eighty thousand years is a long time to keep inquisitive human beings happy; it encompasses the entire civilized span of mankind, plus plenty more. The riots, rebellions and wars that have plagued Earth in one-thousandth of that time do not augur well for multi-generation star travel. Bernal recognized this drawback himself when, in *The World, The Flesh, and The Devil*, he wrote that such journeys 'would require a self-sacrifice and a perfection of educational method that we could hardly demand at the present'.

The immense distances involved in space flight are made clearer by a familiar trick. We put together a model, either real or imaginary, to represent the Solar System. With the Sun as a football or a nine-inch balloon, the contents of the Solar System would fit into a small en-

Data charts

	Miles		Million miles		Million million miles
Local walk	0.25	To Moon	0.25	Comet (aphelion)	0.25
Between nearby cities	25	Venus (nearest)	25	Proxima Centauri	25
Between major cities	100	To Sun	93	Tau Ceti	70
Intercontinental flight	4,000	To edge of solar system	3,600	Betelgeuse	3,800

velope. The Earth would be the size of pinhead, and even Jupiter would rate no bigger than a table-tennis ball. But the Earth-Sun distance on this scale would be 80 feet, and we would have to walk half a mile before we came to the orbit of the outermost planet. The gulf from the Earth to the Moon—so far the only step into space that men have taken—would be 2½ inches.

To reach the nearest star we would have to walk 4,000 miles.

We can look at it another way. Comparing the values in the chart, we can see that to match familiar distances on Earth with distances in the Solar System, we have to scale up by a million times. And to compare interplanetary distances with the stretch to the nearest stars, we have to scale up by a million times again. The dawn of the space age has meant that most people are slowly coming to grips with the first of those changes of scale. But the second million fold scaling up still daunts even the most daring of space-age thinkers.

One way that we might cheat the immense time-spans to the stars is by deep-freezing the astronauts so that their bodies remain in suspended animation while a machine pilots the craft and—we hope—thaws the crew out again at journey's end.

This technique, beloved of science-fiction writers, has been seriously considered by scientists. Physiologist B.M. Gooden of the University of Adelaide dwelt on some of its possibilities in an article in *Spaceflight* in 1970. Pointing out that nearly 4 kilogrammes of food, water and oxygen are needed for each astronaut per day in space, he noted that there are two ways of reducing the quantity of supplies needed for a flight. One is to have a regenerative system that recycles all waste products—and, indeed, just such a system was given a long-term trial by the Soviet Union in 1968, when three men spent a year in a closed environment. In 1973, four men lived for six months in an artificial 'space cabin', using plants for food and to supply air. In an experiment in 1974, one man spent a month in a sealed cabin with chlorella cells to recycle air and water.

Gooden's second method was to reduce the bodily needs of crew members during the flight, so that they did not burn up consumables unnecessarily. He referred to the ability of some animals, including certain mammals, to hibernate for the winter. In this state, their body temperature drops to only a little above that of their surroundings, so that energy is not wasted in producing unnecessary body heat. During hibernation, the body's metabolic rate drops dramatically, and the rates of both heartbeat and breathing are drastically reduced—a condition very similar to that in some meditative trances.

Hibernation happens naturally. But there is a medical practice, called hypothermia, in which the body tissues are artificially chilled to arrest their metabolism. This is used to prevent damage to tissues during operations when the blood flow has to be cut back.

By contrast, freezing a human to the ice-point produces ice crystals that destroy the body's tissues. 'The long-term storage of whole animals at freezing point or below followed by 100 per cent reanimation is still well beyond present capabilities,' says Gooden. Even to chill the body to the surgically used temperatures of 10°-15°C would still require artificial-heart equipment because breathing and heartbeat cease at that stage, although cell metabolism continues at a depressed rate.

Most promising, thinks Gooden, would be a technique to enable humans to regulate their own body temperature like hibernators.

We saw in Chapter Three that spacefaring may be an accepted lifestyle for whole communities in the future. So for these people, a long-

The travel times of light and radio waves.

term trip to the stars may be a natural goal to pursue—and far more constructive than an endless circling of the Sun.

Isaac Asimov, in his factually based look into the future called *Our World in Space*, envisaged that colonies on the asteroids would become star travellers. An asteroid such as Ceres, he says, 'in a sense will be all surface.' By slicing it judiciously, we can use its inside as well as its surface for habitation; unlike on a large planet, there is no appreciable gravitational compression of an asteroid's interior.

Asimov foresaw that such an asteroid-ark would embark on an interstellar journey, perhaps hitching up a comet to provide fuel for its engines on the long journey. Eventually it might encounter spacefarers of another civilization, members of an intellectual brotherhood between the stars that at this moment exists without our knowledge. 'As part of this vast brotherhood of intelligence spanning and filling the Universe, mankind might find its true goal at last,' says Asimov.

Once free of the Earth, a space vehicle needs no gargantuan source of power to ease it towards the stars. Escape velocity from the Solar System is only 50 per cent greater than escape velocity from the Earth. In fact, by using the orbital speed of a giant planet such as Jupiter to tug it along, a craft can be flung free of the Solar System without expending even that 50 per cent extra. The Pioneer probes are getting just such a free ride.

The problem comes, though, when we try to cut down that yawning transit time. Chemical rockets hold out no hope, because although they pack a massive initial punch to boost a payload away from the Earth, they are just too heavy to be any help in speeding our flight to the stars. The 3,000 tons of a Saturn 5's fuel are all burnt in 12 minutes. To reach any appreciable fraction of the speed of light—a necessity if we are to make it to the stars in a lifetime—we would need to keep the engines turning for weeks, months or even years.

Clearly, we need a new source of power for effective interstellar flight. There are plenty of ideas, but few practical results.

One idea that is perfectly workable in theory is to use the weak pressure of sunlight to gently 'blow' a giant reflective sail with a cargo attached, like a clipper of the cosmos.

The pressure of light has clearly visible effects. The streaming tails of comets are propelled by the outflow of light and the solar 'wind' of atomic particles. Together, these could combine to edge a cosmic clipper into deepest space at ever-increasing speed, accelerating by perhaps an inch each second. The power of sunlight has the advantage that it operates continuously, thus building up a high final speed. And it is free.

'A vehicle can be accelerated almost to the speed of light if an emitter on the Earth can accurately project light onto its mirror,' noted Hungarian physicist Professor G. Marx in a paper published in *Nature* in 1966. He was envisaging a laser-propelled spacecraft that would, in effect, have its propulsion system left behind on Earth.

The efficiency of such a system rises to 100 per cent as the speed of light is approached, he said. In theory, of course, laser propulsion removes all the principal difficulties of interstellar flight. But, in practice, Marx agreed that building such a system would be impossible 'in the next few decades'.

One drawback that he quickly pointed out was that the craft's maximum range of operation depends on the laser radiating-area, the craft's reflecting mirror area, and the energy of the light-beam. Professor Marx calculated that to reach a 0.1-light-year range would require a radiating-area of one square kilometre emitting hard X rays, with a suitable mirror of area several square kilometres on the craft.

Such a scheme would solve the problem of the craft's acceleration—but not deceleration. However, even that factor is taken into account in the proposal of a Canadian engineer, Philip Norem, for a 1,000-ton interstellar galleon, propelled towards the stars by a laser array on Earth.

The craft, attached to a giant sail by cables 20 miles long, would coast outwards at a speed of 62,500 miles per second for 8½ years before veering off course for around 20 years, turning into the beam to use it as a brake for another 8½ years. This would bring the craft to α Centauri.

Setting up a similar laser system there would cut the return-trip time to 10 years, Norem calculates.

Such an arrangement is a variant on the photon rocket, a legendary propulsion unit (which is beyond our capability to build) that would eject a beam of light to propel the payload. A photon rocket would have the advantage that, although its thrust at any instant is low, the exhaust speed is the fastest that can possibly be attained. Given a long enough firing time, a photon-propelled rocket would build up to the speed of its exhaust—the speed of light.

Unfortunately, the engineering problems associated with generating sufficient energy to create suitable photon thrust, and of reflecting this powerful beam out of the ship, seem so insuperable by today's technology that we must regretfully shelve any ideas of photon propulsion—at least for the time being.

One alternative propulsive force that has actually been tried on small satellites in space is the electric rocket, or ion drive. In this, the atoms of the fuel are ionized (have their electrons stripped away) by heat, and the ionized gas is accelerated by electric fields. This can give very high exhaust velocities, but again the thrust is low.

A more promising development for interstellar flight is the nuclear rocket, variants of which were being built and tested in the American Nerva programme until cut by a swing of the budgetary axe. As in a nuclear-powered vessel at sea, the nuclear reactor serves as a particularly efficient way of generating heat. At the moment, only a fission-powered reactor is technologically available; controlled fusion power is some way off yet, though engineers are attempting to develop it for power stations on Earth.

In its simplest form, the nuclear rocket uses as a propellant liquid hydrogen, which is heated to a gas by the reactor and expelled at high speed. The advantage of nuclear power, of course, is that its energy release is so much more efficient than the chemical reactions that power conventional rockets. What's more, only the liquid-hydrogen propellant needs to be carried. In chemical rockets, an oxidizer is also needed to create the chemical burning that produces the hot exhaust gas.

Perhaps the best variant of the nuclear rocket would be a hybrid incorporating electrical propulsion, in which the nuclear power would be used to ionize the propellant and to generate the electric field that accelerates it. A nuclear-electric rocket is within the grasp of present-day technology if we are willing to spend the money on it. But its flight time to even the nearby stars would still be of the order of centuries. And fully assembled it would weigh as much as several hundred Saturn 5s.

For a round trip, the problem is correspondingly increased. Surveying the field of interstellar rocketry, propulsion engineer Alan Bond has said, 'It seems certain that no vehicle built with today's know-how will ever return from a star mission.'

Although controlled fusion, of the hydrogen-to-helium type that powers the Sun, has not yet been perfected for use on Earth, we may still be able to use it for rocket propulsion. The fusion reaction takes place in a gas so hot that it would melt any metal container. Therefore, it has to be contained by magnetic fields—but it still tends to break free. In a rocket, of course, this would actually be a desirable feature, because some propellant must be expelled to push the craft along. So the weakness of fusion technology for terrestrial use is no drawback for rocket propulsion.

An even more promising approach to the problem may be to trigger tiny fusion explosions with laser-beams, so that the craft is propelled by pulses of high energy. The fusionable material would be carried in small pellets, each of which might release power equivalent to 10 tons of T.N.T. In such a ship, the fuel would weigh much less in proportion to the payload than a fission reactor, such as Nerva.

No one will need to be told that there is a more explosive way of releasing nuclear energy than in a reactor. Literally bombing into space has been considered by several investigators, and not least in a secret U.S. study named Project Orion.

A series of nuclear blasts directly behind a spacecraft can provide a powerful kick. Hydrogen fusion, as in an H-bomb, is the most efficient energy-generating source currently available to us. Releasing power in such a way may not be the most elegant of engineering solutions, but it is the best way that we have, and therefore worthy of further consideration. We shall look in more detail at one design-study at the end of this chapter.

What other ways can we think of for generating energy? One wild suggestion is merging matter and anti-matter. This, in theory, releases energy with an efficiency of 100 per cent—matter and anti-matter annihilate each other on meeting. But such a speculation, while theoretically sound, remains for all practical purposes in the realms of science fiction.

So too do hopes of using the weak gravitational or magnetic fields in space, or even of modifying space itself by means of the mythical 'warp drives' of pulp-novel starfarers.

In fact, the extreme requirements of enormous power, extended acceleration, and limited time for the journey have led many scientists to the conclusion that star travel is a practical impossibility. Sebastian von Hoerner, examining the prospects for personal contact between civilizations, drew this conclusion: 'Space travel, even in the most distant future, will be confined completely to our own planetary system.'

One daring proposal that would overcome at least some of these handicaps is the interstellar ramjet, which received enthusiastic support when it was propounded by rocket engineer Robert Bussard in 1960.* The ramjet is attractive in that, unlike most rockets, it actually thrives on long-distance travel. What's more, its fuel is free—because the proposal is to scoop up hydrogen gas from the thin clouds in the Galaxy and feed it into a fusion reactor.

The ramjet has the great advantage that it could reach accelerations close to the pull of gravity on Earth (1 g), which is not only conspicuously better than any rival starship design but will also feel completely natural to the crew.

The actual rate of acceleration depends on the density of the

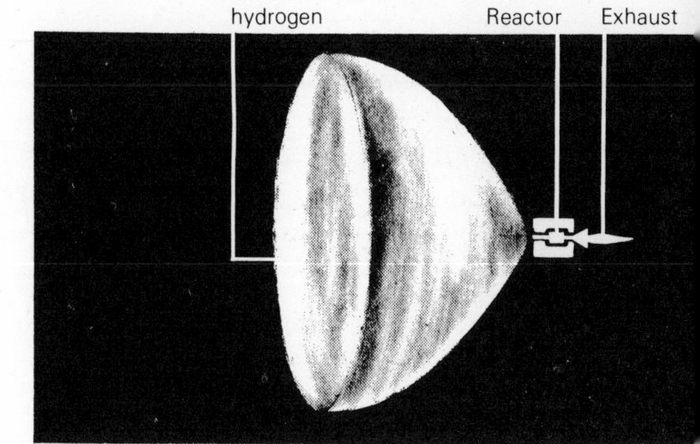

Sketch of an imaginary ramjet vehicle.

* In *Astronautica Acta*, 6, 179 (1960).

hydrogen clouds, and the frontal area of the ship—the ramscoop. For high accelerations, Bussard concluded, 'Interstellar ramjet ships must be large in size and relatively tenuous in construction.' That 'tenuous construction' means not a solid scoop but an electric or magnetic deflector. In the hypothetical ship that Bussard visualized, the ionized hydrogen gas would be deflected by these fields to a focus at the fusion reactor. Bussard's own words underlined the speculative nature of the system that he was proposing:

> At the focal point these ions are led into a fusion reactor of unspecified (indeed, unknown) type, made to react and generate power which is then fed back into the fusion products through a similarly unspecified conversion device, to the increase of their kinetic energy and momentum, with consequent reaction on and acceleration of the vehicle.

The energy required to maintain the electric or magnetic deflection field would be only a tiny part of the fusion reactor's output. To start the whole machine off it would need an initial push. But, said Bussard, 'Boosting to velocities readily reached by present-day chemical rockets would be sufficient for any desired interstellar flight.'

Bussard found that a vehicle of 1,000 tons mass would require an intake width of over 100 kilometres for even high-density clouds. 'This is very large by ordinary standards,' he confessed, adding, 'but then, on any account, interstellar travel is inherently a rather grand undertaking.'

However he modified his statement by underlining one fact: 'There is no thought that anything resembling the required reactor and propulsion system could be built today.' Though he added more hopefully, 'There is likewise no reason to assume such a device is forever impossible.'

One man who has criticized the limitations of the interstellar ramjet is aeronautical engineer Dr Anthony Martin of London's City University. Problems of the ramjet are that the particles that are trapped in the scoop's magnetic field will radiate away some of their energy before they can be used, and there will also be incomplete fusion of all the material scooped up. These losses grow as the vehicle speeds up, until they eventually prevent any further acceleration. 'The vehicle cannot approach the velocity of light as closely as one would like,' Martin concluded.

Even assuming that a sufficiently stable magnetic field could be generated round the ramjet to scoop in fuel, calculations show that the field strength must increase as the ship's speed increases. Yet these magnetic forces will also tend to burst the ramjet's physical structure, and so there comes a point when acceleration must ease off to prevent the whole craft breaking up. The stronger the material of the craft, the longer the slowing down can be delayed. 'Perhaps our grandsons will be flying to the stars in diamond ships with magnetic sails,' Dr Martin poetically hazarded.

In letters to the magazine *Spaceflight*, readers pointed out another drawback of the ramjet—its inability to decelerate. Turning the craft round would be pointless because it would not then be able to scoop in any fuel at all. Some way would be needed of reversing the thrust, like in a jet aircraft, although without disturbing the flow of hydrogen into the intake. Another idea was to modify the intake field so that it acted like a magnetic parachute.

For a manned fusion-powered spacecraft, shielding the crew from the radiation of the power plant and the scooped-up particles is a major consideration. As Dr Edward Purcell of Harvard once pointed out, near the speed of light every incoming hydrogen atom looks as though it had been flung at you from a high-powered particle accelerator. So the radiation danger from high-speed star flight presents problems all the way.

In desperate searches to ease our chances of star flight, some speculative writers have referred to possibilities such as faster-than-light travel, perhaps by using the hypothetical tachyon particles that can *only* travel faster than light, or by finding the mythical wormholes in space that may provide an instantaneous passage to another part of the Universe's mysterious fabric.

Pondering such fictional chinks in the light-barrier is fun, but unhelpful. A more practical solution, made at the Byurakan conference by G. Marx, is to have a three-stage system in which the starship starts off with ion propulsion using nuclear fuel, is then accelerated faster by X-ray laser-beam, and finally switches to ramjet propulsion at high speed in dense interstellar clouds.

And deceleration? That would be done with a laser-beam set up by the civilization at the destination. 'Direct contact can be realized only after having established radio contact,' asserts Marx. 'It can be only a mutual undertaking of two friendly and cooperative societies.'

Assuming that we can find a way to accelerate at 1 *g*, after about a year we would be travelling close to the speed of light, relative to our home on Earth. But even then one might doubt the value of this, when faced with the enormity of the Universe. Surely a round trip to the nearest star would take almost nine years? And stars farther away than 25 light years must surely be for ever out of range for a return trip in an astronaut's natural life-span?

This is where the theory of relativity informs us that time on a speeding spaceship does not pass as quickly as it does back here on Earth. In effect, travel close to the speed of light provides us with a natural time machine. If we went fast enough, we could circumnavigate the Universe in a lifetime.

Unfortunately, this form of time travel is strictly one-way. An astronaut who set out to explore the farthest reaches of space in his near-the-speed-of-light ship would never have a home to return to. By the time his wandering was over, the Earth and its Sun would have died.

This apparent absurdity is a natural consequence of the fact that the speed of light is the fastest speed we can aspire to. As we approach this magic limit, some totally unexpected things start to happen that we never notice at the comparatively limited speeds that we are used to on Earth and even in today's spacecraft.

The speed of light presents a barrier that we could never break, even if we expended all the energy in the Universe. As we inch closer to light's speed—as seen from Earth—the mass of the spacecraft and everything in it would slowly increase as a result of absorbing all the extra energy used to accelerate us. And in league with this, a clock on board the ship would seem to slow in comparison with its regularly ticking counterpart on Earth. At the speed of light itself, time stands still.

These predictions of relativity are used in particle-accelerating machines, in which sub-atomic specks of matter are whisked round at speeds close to that of light. Sure enough, the masses of the particles increase and their lifetimes are extended.

To further check on the time effect, two American physicists flew an accurate atomic clock round the globe and then compared it with a stationary counterpart. The two disagreed by tiny but appreciable amounts, as theory predicted. Not only is it the atomic processes in a clock that slow down. So also do the biological processes of living-beings: suspended animation without freezing.

But the crew of a high-speed starship would notice nothing different during their journey. Everything would seem to progress normally, for everything would be slowed by the same amount. Only when the travellers returned to Earth would they meet the full and drastic consequences of the difference in time . . .

Discussing this so-called phenomenon of time dilation, mathematician Derek Lawden of the University of Aston pointed out that if the astronauts were turned into radiation, by some unspecified process, they could be transported across space in what would seem to them to be no time at all. This raises visions of the 'matter transporters' of science-fiction.

But let us demand fewer miracles from the engineers of the future and imagine instead that we are able to develop a rocket that simply accelerates at 1 g for long periods of time. What does relativity predict about the outcome? Sebastian von Hoerner has prepared the accompanying chart which compares the round-trip times for various distances, assuming that the ship spends half of its time accelerating at 1 g and then decelerating at the same rate on each leg of the journey. These results are either exhilarating, or they batten down the lid on star travel for ever, depending on your attitude.

One man who is not depressed by these results is, predictably, Carl Sagan. In a classic paper—with the daunting title of Direct Contact Among Galactic Civilizations by Relativistic Interstellar Spaceflight*—Sagan argued that interstellar flight has 'several obvious advantages' as a means of contact between civilizations hundreds of light years apart. Radio would not allow communication over such a distance inside a human lifetime, whereas high-speed starflight would.

Sagan pointed out that with uniform 1 *g* acceleration and deceleration, 'all points in the Galaxy are accessible within the lifetime of a human crew.' And he drew particular attention to the ramjet as a means of achieving such an aim. Sketching the progress of such journeys on a graph, he showed that we could reach the galactic centre in 21 years, and fly to the Andromeda galaxy in 28 years. Inhabitants of higher-gravity planets than Earth might choose accelerations of 2 g or more, leading to a halving of even these short times.

Sagan's paper, written on a grand and uncompromising scale, was one of the early bold forays in this field. He said, 'Allowing for a modicum of scientific and technological progress within the next few centuries, I believe that interstellar spaceflight at relativistic velocities to the farthest reaches of our Galaxy is a feasible objective for humanity.' And if *we* can do it, he pointed out, so can others; thus a planet such as our own may expect to be visited every few thousand years.

Perhaps a flight round the Galaxy, or beyond, is expecting too high a performance from our hypothetical starship's engines. Let us set our sights a little lower, and look at some of the targets we might aim for in the locality of the Sun.

Aerospace engineer James Strong plotted the accompanying view of stars we might wish to visit within 20 light years of the Sun. He found

* *Planetary and Space Science*, II, 485-98.

Total duration and distance reached, with constant acceleration and deceleration at 1 *g*

Duration (out and back) years		
For crew on board rocket	For people on Earth	Distance reached, light years
1	1.0	0.06
2	2.1	0.25
5	6.5	1.7
10	24	9.8
15	80	37
20	270	137
25	910	460
30	3,100	1,570
40	36,000	17,600
50	420,000	210,000
60	5,000,000	2,500,000

(after Sebastian von Hoerner)

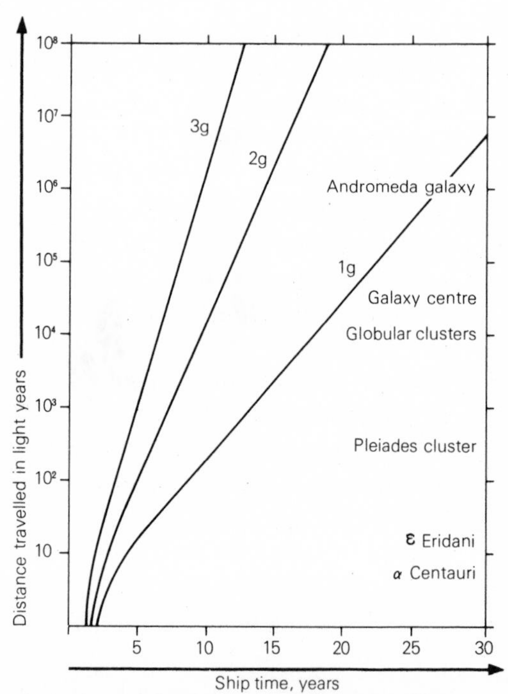

Sketch graph of the time taken to reach various targets aboard a spaceship moving at very high velocity. (*After Carl Sagan*)

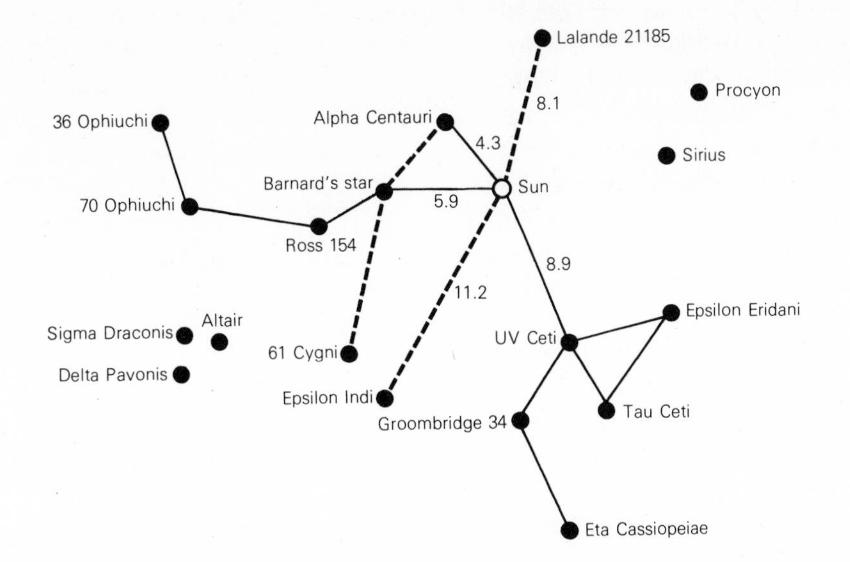

Chart of the Sun and nearby stars, showing their tendency to align in two preferred directions either side of the Sun. Distances are in light years. (*After James Strong*).

Direct distances to other stars named:	
Sirius	8.7
Ross 154	9.5
Epsilon Eridani	10.7
61 Cygni	11.2
Procyon	11.4
Groombridge 34	11.6
Tau Ceti	11.9
Altair	16.0
70 Ophiuchi	16.4
Eta Cassiopeiae	18.0
Sigma Draconis	18.2
36 Ophiuchi	18.2
Delta Pavonis	19.2

that, it we are to make our way into the Galaxy by hopping from one star to another, then two main directions show up clearly: towards Tau Ceti, Epsilon Eridani and beyond; or past Barnard's star. 'The sole merit of journeying to Alpha Centauri in the first instance is that of nearness,' he said. 'But this is the only reason, for beyond stretches a yawning gap of 15.8 light years before the next star of any consequence is encountered.'

On our hops in search of stars with planets, we might wish to use red dwarf stars such as Ross 154 or UV Ceti as stepping-stones. But these should be approached with care, says Strong, because they are subject to sudden flares of intense radiation.

Some of the hops that Strong charts are quite long, and he advocates prior reconnaissance by automatic probes. 'One cannot afford to make mistakes on stellar adventures of this scale,' he warns.

Looking to wider horizons, the alignment of Sun-like stars seems to continue farther afield, possibly as a result of the Galaxy's spiral-arm structure. If radio signals are being beamed to us, Strong suggests, 'the odds are slightly more favourable that they would come in along this axis.'

Strong then goes on to propose that, if galactic communities exist, they would also tend to spread towards stars like their own sun. Therefore, as we star-hop in the future, we might encounter such a race engaged in similar expansion coming the other way.

What would we see from our ship during a high-speed dash to the stars? Various unexpected visual effects would become apparent. One is that the very shape of the Universe would seem to be distorted by our rapid flight. The stars would appear to cluster ahead of us, in an effect similar to that in a moving vehicle when rain seems to come from in front. An ordinary flat chart, therefore, would become increasingly inaccurate.

Additionally, as we move out into the Galaxy the familiar perspectives would change. 'The reference background of stars can no longer be regarded as infinitely distant,' says NASA electronics researcher Saul Moskowitz, who has investigated the visual aspects of interstellar flight. We have to take into account the fact that the stars are distributed in three dimensions.

Also, in what is termed the Doppler effect, our high-speed flight will cause a change in the wavelength of light that reaches us. Stars ahead of the spacecraft will seem bluer as the approaching light-waves pile up into shorter wavelengths, and the stars behind (including our Sun) will seem redder as their light stretches out behind us into longer wavelengths. 'An observer aboard such a vehicle will see a spectacularly changing Universe,' says Moskowitz.

Moskowitz fed star positions into a computer to show how the sky's appearance would change as we journeyed in space. Not only positions alter but also brightnesses; as seen from α Centauri, the Sun would be an ordinary-looking star in the famous W-shape of the constellation Cassiopeia.

The Doppler effect introduces another change in brightness, additional to that caused by the altered distances of stars. The peak wavelength of a star's output depends on its temperature. The hottest stars give out most radiation at short wavelengths. They would seem to brighten as we receded, because their short wavelengths would be lengthened by the Doppler effect, thereby entering the visible region. A cool star that gives out peak radiation at long wavelengths would seem to brighten as we approached and its long-wavelength radiation became bunched up into the visual region. The effects would be reversed if our direction of motion was reversed.

Others have described, with some drama, how the Doppler shift would eventually mean that light coming from stars behind and ahead of us would be changed so much in wavelength that the stars would become too faint to be seen. As we accelerated towards the speed of light we would be advancing into a cone of darkness, punctuated only by a few nearby and bright stars. Behind us, a similar cone would grow.

The only distant stars we would see from the ship would lie in a ring round the sky ahead of us. And their light would be Doppler shifted into a rainbow of colours—a starbow. Being unable to see our destination or our place of departure, how would we navigate? Probably, instruments sensitive to the non-visual regions of the spectrum would be able to keep track of our home and target stars. As physicist James Wertz of Moorhead State College in Minnesota has pointed out, the distribution of radiation from red and blue supergiant stars would keep them visible despite the Doppler shifting that consigns less brilliant stars to the cone of darkness. These super-giants could act as navigational beacons.

Yet Wertz points out that our poor knowledge of star distances handicaps our celestial navigation. Three-dimensional computer maps, like that of Saul Moskowitz, become rapidly less accurate as we progress away from the Sun.

Probably we would want a system that involves inertial navigation, which is used increasingly in aviation and shipping. Inertial navigation takes all accelerations and course changes into account to show how the craft's position has changed in relation to its starting-point.

Tony Lawton has outlined a navigational system for starflight that combines celestial observations with an inertial navigation computer. The ship's true speed relative to the stars would be found by observing the size of the starbow and the amount of Doppler shift of light from our Sun. 'The majority of equipment for tomorrow's navigation is almost available from today's technology,' he says.

Assuming that we one day thread our way between the stars, what might we find at our destination?

Even if there are inhabited planets, this fact will not be immediately apparent. Carl Sagan has undertaken searches for signs of life on Earth from space-probe photographs—and found that, at a resolution

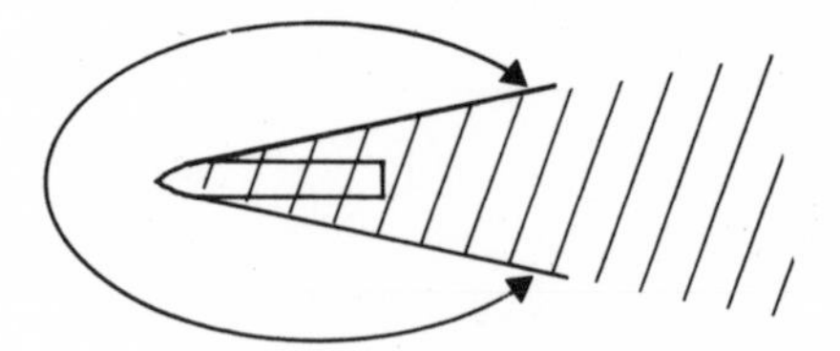

One third the speed of light

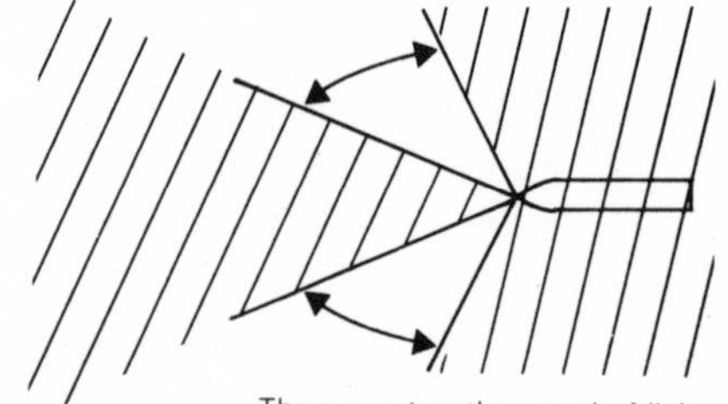

Half the speed of light

Three-quarters the speed of light

View of stars from bridge of starship approaching speed of light.

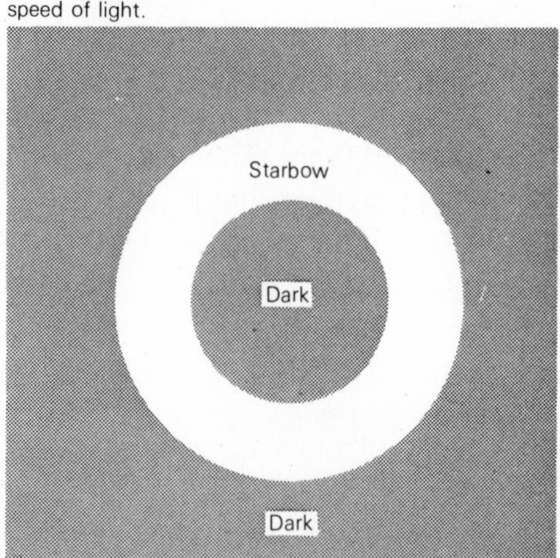

Diminishing areas of visibility from a starship due to effects mentioned in the text

of a few kilometres, there is no indication of human habitation. Only when the resolution is improved to a few hundred metres does the patchwork shape of fields begin to show up.

A more promising approach to finding life relies on a kind of planetary sensor like that used in science-fiction stories and the Star Trek television series. Professor James Lovelock of Reading University has described how we could build such a detector; he calls it a telebioscope. Life on Earth needs certain conditions of temperature, pressure and chemicals. In return, living things change the amounts of different gases in the air—for instance, when plants break down carbon dioxide to release oxygen. By scanning a planet's atmosphere we could therefore tell whether there were living things below.

'A probe with such a device on board approaching our solar system could single out the Earth,' says Professor Lovelock. 'The hardware for building such a thing is available now.'

For Lovelock's vision to come true, all we need is a starship. And propulsion engineer Alan Bond, a former member of Rolls Royce's rocket-engine team, is one man who believes that such a vessel is almost within our grasp. Tired of pessimistic predictions that starflight will for ever be beyond us, he examined the requirements of such a mission. His analysis of rocket-propulsion systems showed that the nuclear-bomb type of rocket would allow us to mount a starflight mission with the sort of technology likely to be available around the year 2000.

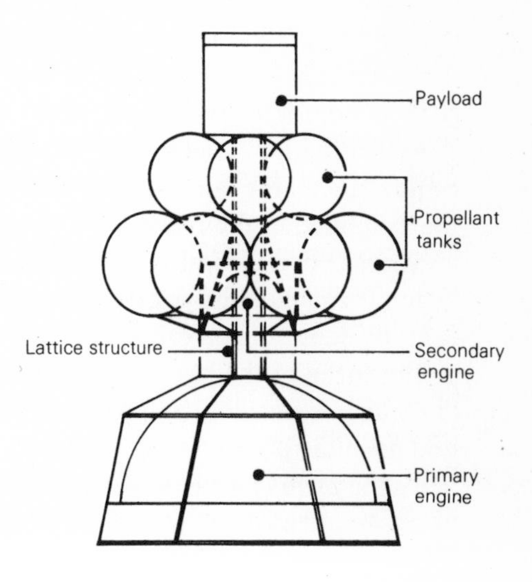

A sketch of the starship being designed by members of the British Interplanetary Society. (*After Alan Bond*)

Bond now heads a study group of the British Interplanetary Society, who are designing an unmanned probe with the planetary system of Barnard's star as its target. Their conclusion: 'Yes, it is possible to fly to a star.' And they have the calculations to prove it.

The type of spacecraft that the eleven-man team is sketching out is a two-stage device, with a total length of about 200 metres. They have decided on a starship with an overall weight of 68,000 tons, carrying a payload of 400 tons—equal to more than 5 of America's Skylab space stations.

Says Bond, 'The vehicle, wherever it lights up from in the Solar System, will be under boost for about 5 years—2½ years for each stage.' The probe will then coast towards its destination at a speed of around 100 million miles per hour; total time for the flight is 48 years. Originally, the team was hoping for a shorter flight so that, should such a mission ever be mounted, it could be completed within the lifetime of the youngest members of the project. However, they found that adding just a few years to the probe's flight time allowed them to halve the amount of propellant it would need to carry.

There are two possible nuclear reactions to power the starship: lithium/hydrogen and deuterium/helium-3. In Bond's opinion, only the latter is within our technology. 'But helium-3 is almost non-existent on Earth,' he says.

One way of getting it is to extract it from Jupiter's atmosphere, which has abundant supplies. 'It means processing an awful lot of Jupiter's atmosphere,' Bond admits. 'Nonetheless, the removal of sufficient helium-3 to power 10 starships should represent no major problem. There's sufficient helium-3 on Jupiter for about 10,000 million starships of this size.'

Another possibility is to breed helium-3 artificially. 'You've got to do that on the Moon,' comments Bond. 'The waste heat liberated by the breeding process is equivalent to the amount of energy that we consume at our present rate in 700 years.'

The scale of the starship means that it will have to be built in orbit,

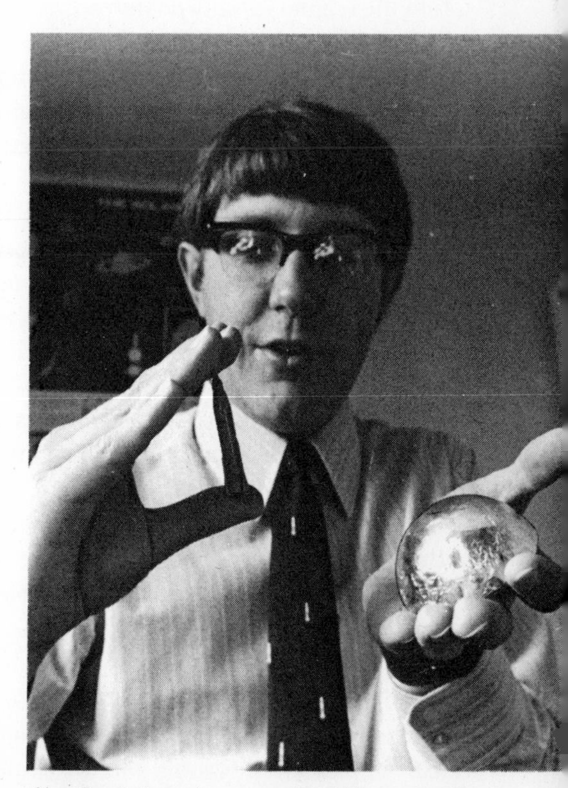
Alan Bond shows how a model of his proposed starship's hemispherical reaction chamber compares in size with a Saturn 5 rocket. (*Ian Ridpath*)

and Bond foresees this being done most economically from a colony on the Moon. Much of the starship's structure will be stainless steel, with the propellant tanks made of aluminium alloys. Molybdenum will be needed for the massive reaction chamber because of the enormous heat generated there.

The first-stage reaction chamber is a hemisphere 100 metres wide—roughly the height of a Saturn 5 rocket—but with a thickness of one millimetre. Bond compares it to a large foil dish.

It is in the reaction chamber that the bombs are ignited, at the staggering rate of 250 per second. Each bomb is a complex structure about half the size of a table-tennis ball, and releases energy equivalent to about 90 tons of T.N.T. The bombs are stored, in their millions, ready-made in the fuel tanks. 'The machinery to get them from the fuel tanks into the engine is more akin to a bottling factory than a rocket,' confesses Bond with north-of-England candour.

A dramatic illustration of the enormity of the engineering requirements for this starship is that each bomb has to be injected into the combustion chamber by a magnetic piston at a speed of twelve kilometres per second—greater than the escape velocity of the Earth. When it reaches the centre of the combustion-chamber dish the bomb is hit by energetic electron beams and ignites explosively.

The force of the blast is absorbed by a strong magnetic field in the chamber, which acts as a kind of magnetic spring to cushion the shock and push the probe along.

As the probe moves towards the stars, initially with an acceleration of one-hundredth the Earth's gravity but eventually at just over 1 g, the empty fuel tanks begin to drop away. The first stage has six tanks, which drop off in groups of two. When the first stage is discarded, at about half the final velocity, the smaller 40-metre reaction chamber of the second stage takes over. This stage has 4 drop tanks. A fifth tank is centrally mounted behind the payload, and remains with the vehicle to fuel the engine for any course adjustments.

The front of the probe is shielded against the impact of particles from space by a 50-ton graphite screen. The payload, and the second stage's permanent fuel tank and engine, shelter behind this 40-metre-wide screen as they coast to Barnard's star at around 15 per cent of the speed of light.

'We're really flying a laboratory from the Sun in the direction of Barnard's star,' says Bond in description of the craft's on-board equipment. During the mission, small sub-probes will be dispatched to examine the surrounding interstellar space. By that stage, the main starship will be running its own affairs with its powerful on-board computer.

Back on Earth, isolated from the probe by the increasingly long travel time of a radio signal, the starship's controllers would sit with large antennae to receive its data. Once the powered section of the flight is over, the design team's communications engineers plan to use the unwanted hemispherical combustion chamber as a radio dish.

As the ship approached Barnard's star, a series of seventeen sub-probes both big and small would be deployed. 'This entire assembly will go through the region around Barnard's star in shotgun fashion, Bond says, spreading his fingers in illustration.

Each of the smaller probes will be independently targeted to take close-up looks at planets and to sample the space round the star. They would relay data back to Earth via the mother ship. Forty years later, the more slowly moving first stage of the rocket would also drift into the Barnard's star system, and might be instrumented to make a second set of readings.

Existing radio dishes, such as that at Arecibo, would be sufficient to communicate with the starship, but we ought to be able to do much better. With a Cyclops-like array we could receive one television frame from the distance of Barnard's star every 3½ hours.

Communications studies for the starship project have been done by Tony Lawton, who says, 'We envisage an on-board storage capability for about 1,000 pictures.' These would be sent in by the sub-probes as they skimmed through the system.

'We think that we could fairly easily see sunspots on Barnard's star. This would itself be an invaluable picture.

'And we may actually see surface features on some of the larger planets surrounding Barnard's star. There would certainly be no transmission problems about getting the data back.'

Assuming that we could mount such a daring mission, when would be a good time to go? Carl Sagan's reply is this: 'The cost of such a mission is very large—something like the cost of the Vietnam war, though obviously a much more worthy enterprise.

'I don't see any way of justifying such expenditures, any more than a manned mission to Mars, in the face of so many pressing alternative social needs.'

Yet with improving technology and the increasing gross national product of the world, the mission begins to get cheaper as time goes on. Sagan says carefully, 'I would think those things begin to make sense around the turn of the century.'

Soviet engineers have also thought, although in less detail, about interstellar probes. At the twenty-fourth International Astronautical Federation congress in Baku, in 1973, applied mathematician Ural Zakirov imagined an advanced civilization in the Barnard's star system sending a probe to investigate us. He assumed constant propulsion from an electric rocket, and said, 'The probe would take about 200 years to travel between the spheres of gravity of Barnard's star and the Sun.'

This is clearly a less dramatic vision than that of Alan Bond. Even so, Bond does not regard his starship study as the best way to reach the stars. Instead, he foresees a modification of the ramjet concept, which he terms the Ram Augmented Interstellar Rocket (RAIR).

This is a rocket/ramjet hybrid, which uses a nuclear power plant to produce thrust, but enhances its performance by scooping up the interstellar medium like a ramjet. The RAIR still has the intake problems of the ordinary ramjet, but the reactor problems, says Bond, fall 'not into foreseeable technology, but perhaps not too far beyond it.'

He has examined the technical details of a mission using such a craft, and says, 'An RAIR may represent a solution to interstellar flight out to perhaps a hundred light years or so until such a time as the full ramjet becomes feasible.'

9

PROBES FROM THE SKIES

Unidentified flying objects, or flying saucers in the looser popular parlance, have been with us in one form or another for a long time. According to some readings, tales of strange aerial phenomena can be found in the Bible and similar ancient texts.

The term 'flying saucer' was coined in 1947 by the popular press when Kenneth Arnold, a private pilot, saw a group of glowing objects that he described as shaped like saucers while flying in Washington state. The popular myth has blossomed from there, with nationwide—some say even worldwide—'flaps' occurring at times of heightened sensitivity. This sensitivity can be public hysteria or genuine aerial activity, depending on your point of view.

Attitudes to U.F.Os vary widely. There are scientists such as Carl Sagan who maintain that 'conscious or unconscious fraud must always be the first suspicion,' or Arthur C. Clarke who says, 'If you've never seen a U.F.O., you're not very observant. And if you've seen as many as I have, you won't believe in them.'

From what has gone before in this book, we can see that there is no reason to deny the possibility of such a visit from the stars. In a lecture in Sydney, Australia, in 1962, Ronald Bracewell, Stanford University's free-thinking radio astronomer, drew a parallel with the voyage of Columbus. Columbus did not find a more advanced civilization in the Americas. 'Had there been one there, it would have discovered Europe,' said Bracewell. His startling conclusion was that advanced communities in space should be here discovering us.

Bracewell is one who believes that radio communications have been going on between civilizations in the Galaxy for a very long time, and that a network of message channels has been set up. 'They are experienced at locating emerging communities such as ours and bringing them into the community,' he said.

This is one very good reason why we should be prepared to listen out for communications from other civilizations. Sebastian von Hoerner has calculated that, if we really are an average civilization, the first interstellar radio contacts in our Galaxy would have been made 5,000 million years ago, before our Sun was even born.

What are the chances that we have been visited by a search party from an alien planet? In *Spaceflight* magazine for December 1972, G.V. Foster attempted such a calculation. He starts by pointing out that the traffic pattern of stars round the Galaxy would have brought a wide selection of stars close to the Sun during the past 4,500 million years. In its history, the Sun is calculated to have circled the Galaxy about 20 times.

Tabulating the number of stars for a given volume around the Sun, Foster finds that within a 50-light-year radius there are on average just over 1,000 stars present, most of which will have moved away again after a million years, to be replaced by other stars. Although there are about 10 stars within 10 light years of the Sun at present, Foster calculates that a total of about 192,000 will have come inside this distance during the Sun's history, some occasionally closing to less than 1 light year. Within a 50-light-year radius, over 4 million stars will have passed during the history of the Solar System.

'The chances of a visitation will be in direct proportion to the number of separate communities in existence in the Galaxy,' points out Foster. Choosing 10 million as the 'best reasonable estimate' of the number of starfaring civilizations in the Galaxy, and making the simplifying assumption that all have existed during the Solar System's history, Foster finds it probable that the Solar System has been visited 20 times by starfarers from within 10 light years.

Of course, if the average travel distance of the starfaring civilizations is greater than this, the number of visits goes up accordingly. For a 50-light-year travel radius the likely number of visits rises to 420. 'Conceivably the solar system could be well documented and even a well-trampled place,' says Foster.

Conceding that his calculations are 'somewhat oversimplified and idealistic', Foster nevertheless feels it worthwhile to speculate upon their implications. For instance, if the remarkable figure of 420 visitations is right—and some may consider it far too optimistic—this still means an average of only one alien expedition about every 10 million years. It's certainly not enough to satisfy the U.F.O. enthusiasts.

Assuming that alien expeditions have come winging their way among the planets we might expect them to have left some kind of visiting card, perhaps in the form of a small commemorative marker as we have attached to the Pioneer probes. Erosion would long ago have worn away any traces of a visit on certain planets, such as Earth. But on virtually airless and erosion-free bodies such as the Moon, Mercury, the satellites of the major planets, and perhaps Mars, we may find startling evidence of extraterrestrial life.

One of the bodies in the Solar System may bear the footprints of an alien space crew, or the discarded equipment of a landing party. The satellites of the planet Jupiter provide the largest surface area on which such artifacts would remain uneroded for long periods.

The innermost or the largest planet in the Solar System may be tagged in some way, thinks Foster. Such a marker might contain a dating system to reveal when it was left, and an inscription like the one on the Pioneer plaques to show where its makers came from.

'The planet Mercury or one of Jupiter's moons may harbour one or more such objects,' says Foster. His views echo the events of the film *2001*, in which Earth astronauts find artifacts of another civilization as they explore the Solar System.

'Artifact devices may well embody the techniques and principles of superhuman knowledge,' says Foster. 'Sooner or later we will almost certainly encounter these objects.'

Possibly the visitors would leave small satellites in orbit round the Sun. But because of the vastness of space these would be difficult to detect. More obvious would be artificial satellites round a planet. There could be a radio transmitter at work in the Solar System, beaming back information to civilizations of distant stars.

In the September/October 1974 issue of *Mercury*, the magazine of the Astronomical Society of the Pacific, mathematician Claude Anderson of the University of California, Berkeley, suggested that visitors from space might leave a reflecting device in orbit round a planet of our Solar System or one of its moons, to give a brilliant and unmistakably artificial reflection to radar beams pointed its way, and he recommended an examination of the outer planets for just such radar-bright spots.

But why send a manned expedition when a probe could come here on its own? This question was first posed by Ronald Bracewell at the same time as Project Ozma was getting under way.* He argued that if communities were on average spread about 100 light years apart, rather than the 10-light-year range of Project Ozma, then radio signals would not be the best means of establishing first contact. Neither the transmitting nor the receiving civilizations would be sure which of the thousand or so available stars to take aim at. Bracewell said:

> The probability that we are listening in their direction at a time when their signals are arriving in our direction clearly works against success. If such a signal were received, the answer would arrive with a 200-year round-trip delay, at least, which seems a very precarious way of initiating relations.*

Instead, he suggested that the civilization trying to attract attention would send out pilot probes to the stars. These would save both sides a great deal of wasted time in signalling and listening. When such an automaton reached another solar system it would put itself into orbit round the distant star and then start to listen for radio transmissions that would provide evidence of technological life.

The probe would flash this message back home for the benefit of its senders, but it would also be capable of doing much more. It would start to feed us information about the distant civilization without our having to wait for the 200-year round-trip time. Said Bracewell:

> A probe encountered at stellar distances from its place of origin may be expected to be packed with information and to be capable of reacting intelligently to interrogation.
>
> One might be in our system now, and if this is the case we should be very careful not to overlook unexplained radio signals that may be received.

Bracewell proposed that the probe would make its presence known by playing back some of the transmissions that it intercepted, thus giving a kind of echo effect to the sender. As an example of what he meant, he referred to mysterious echoes heard in the early days of radio research by the Norwegian meteorologist Professor Carl Störmer, and the Dutch radio engineer Balthasar Van der Pol.

The purpose of the probe's response would be, said Bracewell, 'to attract our attention to the presence of a galactic chain of communities, the nearest of which was so distant that direct first contact by radio seemed hopeless.' It would tell us what frequency to signal on and towards what star. Prophetically, he speculated, 'Should we be surprised if the beginning of its message were a television image of a constellation?'

If the distances between communities is even more than 100 light years, Bracewell believes that contacts will be made in space as probe encounters probe: an interstellar meeting of minds will be between mechanical brains. In Bracewell's visualization: 'I can see these probes being launched in all directions from the parent planets, making occasional contacts, reporting back home, until ultimately the home planets are in direct communication.'

One young man influenced by Bracewell's ideas was the Scottish science-fiction writer Duncan Lunan. In 1972, after a private symposium with friends about interstellar flight, he took a fresh look at the mysterious echoes of the 1920s referred to by Bracewell. Radio pulses

* 1962 Sydney lecture, reprinted in *Interstellar Communication*, edited by A.G.W. Cameron (W.A. Benjamin, California, 1963).

had been sent out by the Philips company at Eindhoven, Holland, during tests for what was to become Radio Hilversum, and were received after delays of many seconds at a station in Oslo, Norway. If the pulses had been reflected directly back from the Earth's ionosphere, they would have arrived after only a fraction of a second.

Upper-atmosphere researchers made several investigations of long-delayed echoes in 1928 and 1929, but no explanation was found. The radio engineers were particularly puzzled by the fact that the delay times of the echoes ranged from a few seconds—which could have been reflections from the Moon—to 32 seconds.

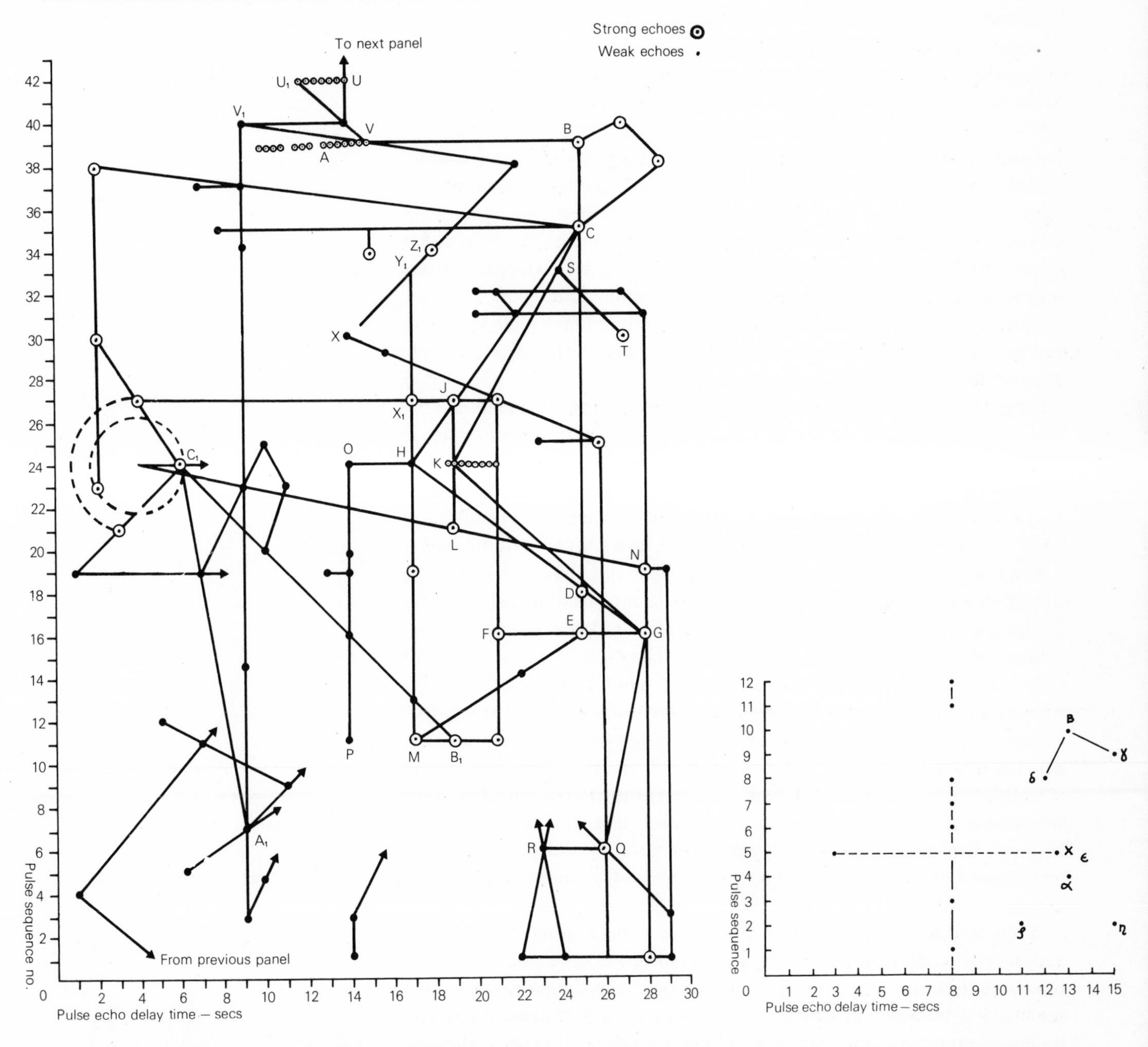

Duncan Lunan's plots of some of the long-delayed echoes received in 1928. The first, small diagram he interpreted as a fragmentary map of the constellation Boötes, with the star Epsilon deliberately misplaced (*to the left*) to draw attention to it. The larger diagram is one of his major so-called panels, and the one which he described at Caxton Hall in 1973. It does not appear as spectacular as the message he was able to draw from it.

Previous attempts at analysing these pulse trains had revealed nothing. But Duncan Lunan was able to interpret them as re-transmissions from a space probe, with the differing delay times forming a code containing a message.

Lunan wrote about his initial researches in the April 1973 issue of *Spaceflight*. Freely conceding that his work was 'a step into the realms of speculation', Lunan showed how a sequence of erratic echoes received by Professor Carl Störmer at Oslo on October 24th, 1928, could be plotted to give a star map of the constellation of Boötes. Other chains of echoes filled in certain of the stars in the region around Boötes.

Boötes, the herdsman, is not one of the most famous constellations. It lies sandwiched between the more familiar outlines of Hercules and Virgo. But its main star, Arcturus, is one of the brightest in the sky. Arcturus had been slightly misplaced on those first maps. The star turns out to have a noticeable movement in the sky, as a result of its movement round the Galaxy, over thousands of years. Lunan suggested that the map showed Arcturus at the position it was in thirteen thousand years ago.

He speculated that that was when the alien probe arrived here and made a map of its home constellation, and then shut down to await signs of life on Earth. When it detected intelligent radio transmissions, the probe's systems were reactivated. It first transmitted its out-of-date star map, but the scientists were unable to decode the message at the time and so failed to acknowledge it. If they had done so, Lunan mused, 'one wonders what other systems might have proved ready for use.'

One star on the supposed maps seems to be continually drawn attention to by reference lines. That star is Epsilon Boötis. Lunan believed that the messages were trying to show that Epsilon Boötis was the home of the probe's makers. Excitingly enough, Lunan found that star to be 103 light years distant—in staggering accord with Bracewell's idea.*

Long chains of delayed echoes were heard by French scientists on May 8th, 9th, and 10th, 1929, in response to transmitted radio pulses. Lunan plotted these and attempted to interpret them.

Again he found a star map of Boötes, but this time with Arcturus in the right position. He wondered if the probe's systems had come back into use to update the old star map of their home. In something like a scene from a science-fiction novel, Lunan, a tall, bony Scot with a straggly beard, presented his conclusions to a packed meeting of the British Interplanetary Society at Caxton Hall, London, in March 1973.

Lunan explained how he had split the transmissions into what he called 'panels'. In the hushed and darkened meeting-hall, he showed a slide of one such panel. It consisted of several rows of dots spaced at irregular intervals. Lunan had drawn lines across the panel to show how the dots link up.

Speaking in measured tones, without the least hint of sensationalism, Lunan described his interpretation of this panel. He believed it to be a picture-story which explains something about the probe's makers. Part of the dot sequence he interpreted to mean: 'Our home is the double star Epsilon Boötis. We live on the sixth planet of seven from the main star. Our probe is in the orbit of your Moon.'

In his book *Man and the Stars*, published in 1974,† Lunan revealed the content of eleven panels, from which he draws the conclusion that the probe's makers had already had to move from the second planet of their system to the sixth, because their star was swelling up at the end of its life and beginning to roast its family of planets. The probe therefore became to Lunan not an exercise in interstellar communication, but a desperate attempt at survival. The inhabitants of Epsilon Boötis were desperately looking for a new home, and perhaps the probe contained the seeds to start a new colony.

*However, a check with more accurate star catalogues has since revealed that Epsilon Boötis is 233 light years away. Lunan recognized that it is a double star; apparently, though, it may actually be triple.

† Souvenir Press, London.

But was all this a little too good to be true? For one thing, a double (or triple) star such as Epsilon Boötis is not supposed to have planets, as we have seen. And also, the dot sequences were so fragmentary and poorly recorded that the smallest errors could upset the entire interpretation.

At the Caxton Hall meeting, Tony Lawton listed several possible natural explanations of the echoes, and emphasized that we must 'Check, check and check again,' before accepting the space-probe theory of the echoes. In a *Spaceflight* article in April 1973, he wrote about the interpretation of various sorts of signals from space, and drew attention to the fact that long-delayed echoes were still received occasionally by radio operators.

Correspondence that followed varied from the enthusiastic to the highly critical. One astronomer scathingly called it 'one of the most remarkable examples of the manipulation of data that I have come across.'

Bracewell himself discussed Lunan's work in the American magazine *Astronautics and Aeronautics*. He pointed out that acknowledgments sent to a messenger probe would assure it that its transmissions were not being wasted. But he added that to receive an acknowledgment, the probe's first message would need to be absolutely convincing. Looking at Lunan's data, he declared: 'This message fails the test of being convincing.'

Lunan countered, saying, 'Even if no probe exists, the discussions and experiments stand as a useful worked example in the logic of interstellar communications.'

Inadvertently, the subject received another boost at a meeting of the International Astronautical Federation in Baku later in 1973, when it was suggested that the strange emissions being picked up in the Soviet Union by Troitsky and Kardashev were due to a 'galactic sputnik' seeking a suitable planet.

Privately, Tony Lawton, a highly qualified electronics engineer and an amateur radio enthusiast, already had doubts that the situation was as clear-cut as Lunan supposed. So with Sidney Newton, a radio-ham friend, he set up an experiment to test Lunan's ideas.

Lawton borrowed an 18-foot-long satellite-tracking aerial from his employers, E.M.I. Electronics, which he set up in his back garden at Shepperton, Middlesex. The sensitivity of the reception equipment was proved when they picked up signals bounced off the Moon from Stanford University in California.

Ten kilometres away at Twickenham, Sid Newton set up a transmitter using a cross-shaped aerial that he had bought for £13. While Lawton listened, he began to send a characteristic Morse call-sign. But, despite more than a year of trying, they were unable to duplicate the pulse trains that Lunan had interpreted as the probe's message.

Research into the earlier work brought Lawton into contact with R.L.A. Borrow, who had assisted Sir Edward Appleton in early investigations into long-delayed echoes (L.D.Es). As he learnt more about L.D.Es, Lawton came to realize that the received pulse trains reported in the 1920s were not continuous, but faded in and out with changing atmospheric conditions.

Most damaging of all to Lunan's case was the discovery of hitherto unpublished records that did not fit into the star-map interpretation at all.

When I visited Lawton and Newton during their experiments, they explained the difficulties of obtaining L.D.Es because of the vast increase of radio traffic since the 1920s. At the time Van der Pol made

Tony Lawton (*foreground*) and Sid Newton with the aerial used in an unsuccessful attempt to verify the ideas of Duncan Lunan. (*Ian Ridpath*)

Moon
Moon's orbit
Lagrangian point
Lagrangian point
60°
60°
Earth

The Lagrangian points in the Moon's orbit where material might lodge

his transmissions the ionosphere was almost totally man-quiet; when Lawton tuned his receiver to 9.55 MHz, the frequency at which those original echoes had been heard, the loudest sound was music from an Indian radio station.

The Lawton-Newton experiments were carried out over frequencies from 1.8 MHz to 146 MHz using a transmitter power of about 50 watts; yet they recorded very few examples of L.D.Es, and certainly no intelligible response to their call-signs.

'If the probe existed, one would expect it to cotton on to what was happening,' says Lawton. 'L.D.Es seem to be completely random. We are positive that they cannot be reproduced at will.' In description of the ionosphere, he explains, 'It's not the placid reflecting mirror that people tend to think it is. It's a seething, turbulent set of clouds far worse and much more disturbed than thunderstorms.'

This property of the ionosphere is an important key to the explanation of L.D.Es advanced in 1970 by Professor F.W. Crawford of Stanford. He developed a theory based on the type of plasma interactions that occur in cyclotrons and which produce vast amounts of radio interference.

In an ionospheric plasma cloud, the radio wave can be amplified by the energy of electrons from the Van Allen belt. This amplification helps to explain the puzzlingly high strength of the delayed echoes. Different-sized clouds will produce varying degrees of delay. With several plasma clouds in the ionosphere, at heights of around 400 kilometres, a whole train of amplified echoes may emerge from one single pulse. 'This is what we think happened in the early days,' says Lawton.

But such an explanation cannot account for L.D.Es at frequencies above 10 to 15 MHz, which the ionosphere does not reflect. The Lawton-Newton team have observed a possible L.D.E. at 137 MHz. So Tony Lawton appeals to a much longer path-length to turn these high frequencies back to Earth.

He explains, 'When a radio wave is reflected by the ionosphere it isn't a true reflection in the way a mirror behaves. It's a steady degree of bending, almost like a prism effect.'

In a prism, the amount of bending depends on the wavelength; the different bending of each colour produces the rainbow effect of the spectrum. And so it is with the ionosphere: the shorter the wavelength, the less the ionosphere will bend it.

'There comes a time, called the critical frequency, when the ionosphere just lets the wave go straight through,' points out Lawton. 'But if there is a large chunk of ionosphere, even the higher frequencies will start to be bent.'

Analysis of the times when L.D.Es are observed has shown that they are most likely to occur when a point in the Moon's orbit called the trailing Lagrange area is in the observer's sky. This Lagrange area is one of the gravitationally stable regions round the Moon; it lags the Moon in its orbit by 60°. Weakly luminous clouds have been observed in this region by astronomers from time to time.

Says Lawton, 'We think that at times, under very special conditions, the ionosphere actually couples with the Lagrange area.'

In their visualization, meteoric gas and dust, possibly reinforced by solar wind particles, are occasionally swept by the spinning Earth into the trailing Lagrange point. In Lawton's words: 'We think that the Lagrange point temporarily achieves a density similar to the upper ionosphere, and comes actually into contact with the upper ionosphere. If a radio wave happens to enter this area under these conditions it won't know whether it is in the ionosphere or not.'

This gives potential path-lengths of a million kilometres or more. Says Lawton, 'If we now invoke Crawford's theory of plasma interaction in an ionosphere of this size and length you can get some colossal effects.' Of the space-probe theory for L.D.Es, Lawton now says, 'It's a nice idea—but that's as far as it goes. We would now definitely say that all records to date do not support the hypothesis that L.D.Es emanate from an artifact.'

He also attacks the logical basis of the probe idea. He takes the view that a probe would not return terrestrial transmissions in a pulse train, but instead would emit a wide-band signal that couldn't be tuned out. 'The object is to attract attention,' he says. This same analysis goes for the Soviet reports of the galactic sputnik.

As a final nail he hammers home the opinion that an advanced civilization would do most of its telecommunications in the 200 MHz region, rather than the much lower frequencies at which the supposed probe messages are heard.

In a 1974 book with journalist Jack Stoneley,* Lawton developed a speculative scenario of what might happen *if* a space probe of the type Bracewell and Lunan envisaged had tried to make contact. He foresaw how the probe would direct us on to the frequency it wanted to

**Is Anyone Out There?* (Warner Paperback Library, New York).

use—probably in the centimetre band, where it could send television pictures directly, without cumbersome L.D.E. pulses.

Yet, in real life, what would he have done had there really been a probe out there waiting for another chance to signal us—and he had been the one to switch it on? Lawton was ready with an answer.

'We would not naïvely have announced it to anybody. We would have kept it extremely quiet for some time—no people apart from those who could be trusted would have been informed until we were sure that what we were receiving was information from a probe.

'If that had happened we would have informed the government and let them take it from there.

'Once a scientific experiment of this nature succeeds it becomes a matter of straightforward politics.'

Just after Christmas 1969, the American Association for the Advancement of Science (A.A.A.S.) held a symposium in Boston to examine the various aspects of the debate on U.F.O.s. Not surprisingly, news of this symposium created heated opposition among some scientists. One of the most active of the organizers, Carl Sagan, was particularly prominent in defending the symposium, which was intended to be as balanced a look as possible at the vexed subject of aerial phenomena.

The participants included astronomers, physicists, psychiatrists, a psychologist, a sociologist and a member of the press. They met in the same year that the controversial Condon report was published, a two-year study of material collected over the preceding 22 years by the U.S. Air Force, mostly under its famous Project Blue Book.

In their report, published as a 967-page book, the Condon committee examine 90 U.F.O. cases, one-third of which they classify as 'unexplained'. Despite this, Condon concluded in his summary: 'Nothing has come from the study of U.F.O.s in the past 21 years that has added to scientific knowledge . . . further extensive study of U.F.O.s probably cannot be justified in the expectation that science will be advanced thereby'—a hint which the U.S. Air Force gladly accepted in December 1969 when it closed its U.F.O. files.

One of the problems with U.F.O. reports is that, as astronomer Thornton Page said at the A.A.A.S. symposium, 'About 90 per cent of the 13,000 reports received by Project Blue Book could have been recognized as normal physical phenomena by persons who had studied elementary astronomy in high school or college.'

Two astronomers at the symposium, William Hartmann and Donald Menzel, were united in their dismissal of the reality of U.F.O.s. Hartmann, of the University of Arizona's lunar and planetary laboratory, took the attitude that close study of U.F.O. cases will always find a rational explanation for the phenomenon seen—the fault lies in poor investigation. He concluded from his involvement on the Condon committee that 'in the U.F.O. business one can trust nothing secondhand.' Having investigated several of the supposedly strongest cases of U.F.O. sightings he states: 'The U.F.O. evidence is very poor.'

Donald Menzel, former director of the Harvard College Observatory, listed over a hundred phenomena that can account for U.F.O. reports, and branded U.F.O.s as 'the Modern Myth'. The possible explanations that Menzel advanced for U.F.O. sightings include natural astronomical and atmospheric phenomena, such as meteors, satellite re-entries, refraction and reflection effects of sunlight, mirages and ball lightning.

People are also frequently fooled, at night especially, by aircraft, weather balloons, searchlights, and even flights of birds. Most baffling of all are purely imaginary effects caused by after-images in the eye of bright lights, imaginary motions of stars in the sky, and even eye defects such as dust specks.

'Man has traditionally tended to construct a myth to explain anything he cannot understand. And this is precisely the way that flying saucers or U.F.O.s came into existence,' declared Menzel. He said he was glad that the Condon report had 'cooled off' interest in U.F.O.s. 'Scientists of the 21st century will look back on U.F.O.s as the greatest nonsense of the 20th century,' he said.

But James McDonald, an atmospheric physicist from the University of Arizona, took the diametrically opposed view that 'something in the nature of extraterrestrial devices engaged in something in the nature of surveillance lies at the heart of the U.F.O. problem.' He charged that all U.F.O. investigations in the past had been inadequate, and believed that science was 'in default' over the handling of the U.F.O. situation.

He made his statement from the experience of having interviewed over five hundred witnesses of U.F.O. cases. 'In my opinion', he said, 'the U.F.O. problem, far from being the "nonsense problem" it has been labelled by many scientists, constitutes an area of extraordinary scientific interest.' And he believed that the Condon committee had done a 'quite inadequate, unsatisfactory' job in investigating many U.F.O. cases.

U.F.O.s are recorded on radar as well as visually. This field was summarized by radar scientist Kenneth Hardy of the U.S. Air Force meteorology laboratory, who reported that clear-air echoes, colloquially termed 'angels', have been detected regularly ever since radar first began to probe the atmosphere. A five-year research programme with ultrasensitive radars at Wallops Island, Virginia, revealed three causes: insects or birds, sudden refractive changes in the atmosphere, and distant objects brought into view by so-called anomalous propagation. 'At no time has any object been detected at Wallops Island which remained unexplained and was therefore put in the category of a U.F.O,' said Hardy.

Atmospheric scientists have found numerous remarkable ways in which atmospheric cells and layers can give false radar targets. Can these account for the mysterious radar U.F.O.s reported?

Unfortunately, U.F.O. photographs are rarely good enough for meaningful analysis, and many are blatant fakes. Engineer Robert Baker has examined cine films purportedly showing U.F.O.s. While he is convinced that they do indeed show 'anomalistic phenomena', the objects appear only as ill-defined blobs and are too indistinct for firm conclusions to be drawn. 'Only two generalizations can be made,' he declares: 'The photographic images usually occur in pairs, and usually exhibit a slightly elliptical form.'

The sociological and psychological aspects of U.F.O.s were also discussed at the symposium by specialists, who examined the roles of rumour, hysteria and human-belief systems in the reporting and analyses of U.F.O.s.

Chicago sociologist Robert Hall was one who was impressed by the apparent certainty among U.F.O. reporters of what they had seen, despite the fact that their observations jarred their own beliefs. Investigators who assert interpretations that conflict with such evidence 'show a startling degree of disrespect for the reason and common sense of intelligent witnesses,' Hall said. 'Hysterical contagion contributes some cases to the massive number of reports,' he believes, but he finds many factors, including the corroboration of reports by independent witnesses, that argue against this as a catch-all explanation. He compared scientists' reactions to U.F.O. reports with the disbelief that greeted the news of Galileo's discoveries through his telescope, and of stones falling from the sky. 'Or do the behavioural scientists have to accept the puzzling and anomalous fact that hundreds of intelligent, responsible witnesses *can* continue to be wrong for many years?' he asked.

In a joint paper, psychiatrists Lester Grinspoon and Alan Persky of Harvard Medical School discussed the mental processes of individuals in relation to U.F.O. reports. They argued that the unconscious processes which occur in all people often express themselves in the form of symbols. Major symbols are the breast and penis, which tally very well with the usual descriptions of U.F.O.s as saucer-shaped or cigar-shaped. 'These considerations may also help to explain some of the emotionalism which surrounds the subject,' Grinspoon and Persky suggested. In a later aside, Carl Sagan pointed out the conflicting fact that the cigar-shaped objects are usually the ones that U.F.O. experts refer to as the mother ships.

In a review paper, Philip Morrison declared, 'I have now, after a couple of years of fairly systematic listening and reading, no sympathy left for the extraterrestrial hypothesis.' Neither were any of the other participants, even if they were impressed by the observational evidence, willing to join James McDonald in his espousal of the idea that U.F.O.s do represent evidence of craft from other worlds.

In another review, Carl Sagan concluded that there may be *too many* U.F.O.s for the extraterrestrial hypothesis to be valid. Assuming that there are one million technical civilizations in the Galaxy, all engaged in starflight, how often would we expect to be visited? As there must be something like 10,000 million interesting places to visit—a tenth of the total number of stars in the Galaxy—then we would expect one visit every 10,000 years.

But perhaps the higher number of reported U.F.O.s indicates that we are something special. If so, then life cannot be a very common phenomenon in the Galaxy—and thus there would be fewer civilizations to send out starships, and we would expect a much smaller number of U.F.O.s.

Sagan deduces, 'The extraterrestrial hypothesis is in some trouble if we're to imagine that even a smallish fraction of the ten or twenty thousand U.F.O. cases reported in the last 20 to 25 years are interstellar in origin.' And he asks, 'Is it possible that we hear so much of this hypothesis because the idea of extraterrestrial visitation somehow resonates with the spirit of the times in which we live?'

One reputable scientist whose name is closely linked with the serious study of U.F.O.s is J. Allen Hynek, professor of astronomy at Northwestern University, Illinois, and former astronomy consultant to Project Blue Book. His involvement began when the Air Force asked him, as the nearest astronomer available at the time, to assist in the investigation. He describes himself as 'the innocent bystander who got shot'.

Initially, Hynek had scoffed at what he saw as the public's gullibility over U.F.O.s. But his association with Project Blue Book convinced him that, once the obvious misidentifications of natural phenomena were eliminated, there remained a puzzling residue of cases for which there was no ready explanation. Now, he believes that a number of substantiated reports point to 'an aspect or domain of the natural world not yet explored by science'. He claims to have demonstrated that

Allen Hynek – not from up there? (*Ian Ridpath*)

U.F.O.s are not *all* misperceptions or hoaxes, but adds, 'It is not known what U.F.O.s are.'

In his work, Hynek has stuck closely to the observational evidence of U.F.O.s and makes no theories about their nature. He assigns U.F.O. reports two indexes: their *strangeness*, or difficulty of explanation in common-sense terms; and their *probability*, or likelihood that the event actually happened as reported. The second index depends on the number and reliability of witnesses. Using these indexes, Hynek makes a plot of cases that remain puzzling after competent investigation. From the cases reported to Blue Book, Hynek found enough that fell into the area of high strangeness and strong probability to suggest that they 'should command attention and challenge science'.

Hynek, though, does not believe that U.F.O.s are nuts-and-bolts spacecraft from other worlds. 'There are too many things against it,' he says. 'It seems ridiculous that any intelligence would come here from such great distances to do such reportedly stupid things like stop cars and frighten people. And there are far, far too many reports.' Instead, Hynek tends to the belief that the answer to the U.F.O. problem lies in realms of physics that we do not yet understand. He points out that it took until nearly the middle of this century before we knew enough physics to explain what makes the stars shine.

Hynek has support from what he terms the 'invisible college' of serious, trained scientists and engineers who believe that the U.F.O. phenomenon is worthy of closer investigation.

Among them are members of the august American Institute of Aeronautics and Astronautics, which in 1967 set up a U.F.O. subcommittee. The committee published a status report in 1970, agreeing with many other scientists that the Condon report should arouse more study of U.F.O.s rather than quell it: 'We find it difficult to ignore the small residue of well-documented but unexplainable cases which form the hard core of the U.F.O. controversy.'

But about the possibility of U.F.O.s representing spacecraft from other worlds, they say, 'There is not sufficient scientific basis at this time to take a position one way or another.'

In the Institute's magazine *Astronautics and Aeronautics* for May 1974 Peter Sturrock, a highly qualified space scientist and astronomer at Stanford University, revealed the results of a questionnaire sent to members in the San Francisco area. Although he found a wide spectrum of opinion on U.F.O.s, many of the Institute's members were willing to disclose descriptions of U.F.O. phenomena that they had seen. In support of what Hynek had found, Sturrock concluded: 'A sample of scientifically trained persons reports aerial phenomena similar to so-called "U.F.O. reports" '.

By contrast, little official work of any kind is done on U.F.Os in the United Kingdom. Of a total of 1,631 reports received by the Ministry of Defence from 1967 to 1972 (an average of nearly one a day), 203 were classed after investigation as satellites or other space debris, 108 were balloons, 170 were celestial objects, 121 were meteorological or natural phenomena, 750 were aircraft, and 106 had insufficient evidence on which to base a decision. However, this leaves an average of over two cases a month as unidentified. 'The only reason these are looked at is to see if there are any air defence implications,' says an Air Force spokesman.

Members of the invisible college at universities throughout the United States have since joined Hynek in forming a loosely organized Center for U.F.O. Studies. This came into existence in November 1973 after a wave of U.F.O. reports had spread across the U.S.A.

'I just got angry,' Hynek says in explanation of his reaction to this wave which, according to the half-million-dollar burial given by the Condon committee, should not have been happening. 'Everybody thought the subject was dead. And yet the corpse refused to remain buried. Nobody was minding the store, nobody was doing anything about it. There was no place a person could go either to obtain reliable information on the subject or to report an event without fear of ridicule.' Hynek investigated some of the cases reported in 1973 and remained convinced, as he had been by his investigations before, that, whatever people were observing, they were experiencing a genuine event.

Now, the U.F.O. Center has a toll-free hot line on which police throughout the U.S. can phone-in U.F.O. cases reported to them. The switchboard, manned round the clock every day, collects basic information on each sighting which is followed up by a personal visit or a questionnaire from the Center. The Center has published a detailed report on the 1,500 or so known cases from the 1973 flap, which would otherwise have gone undocumented. But the volume of work has not stopped there. On average, one call a night comes over the U.F.O. Center's hot line.

All that the U.F.O. investigators can do at present is to define the limits of the problem. Some of the invisible college members might toy with the idea that U.F.O.s point towards a second reality, or evidence of another dimension. If so, there could be a new physics on the horizon, in which thoughts can be transferred and in which we can travel faster than light—or even through time.

Perhaps with this in mind, Hynek predicts, 'When the long-awaited solution to the U.F.O. problem comes, I believe it will prove to be not merely the next small step in the march of science but a mighty and totally unexpected quantum jump.'

It is impossible to write on such an emotive topic as U.F.O.s without being expected to declare a position. Yet what position can anyone take when very sound scientific minds are so unable to agree?

I remain studiously unconvinced that reports of U.F.O.s are evidence that the Earth is the subject of extraterrestrial surveillance. Were we being surveyed, I suspect that we should either know nothing at all about it—or that it would be only too evident. I believe that we are in danger of confusing something that we do not understand for what we might wish to be true. A study of U.F.O.s may tell us more about human beings than about interstellar flight—though that is not an argument against studying them.

Some, of course, maintain that there is evidence, in mysterious tales that indicate crash-landings of extraterrestrial craft. Yet I am dubious that anyone clever enough to travel to our Solar System would make such an elementary landing error.

Others suggest that governments—notably the U.S.—know the secret of U.F.O.s (or are even responsible for them) but are keeping it quiet. Unfortunately, cases such as Watergate, the Pentagon Papers, and various defence leaks make it all too clear that governments are notably poor in keeping the lid on momentous information which, by definition, is of major public interest.

In conclusion, I believe that we still need concrete evidence, in a scientific sense, before anyone can conclude that we have yet been visited by even one flying saucer.

But even as I write these words, I look uneasily over my shoulder into the blackness of night, aware that the aliens may be on their way.

Specimens of report form used by the U.F.O. Center's telephone hot line and follow-up form sent out by the U.F.O. Center (bottom right).

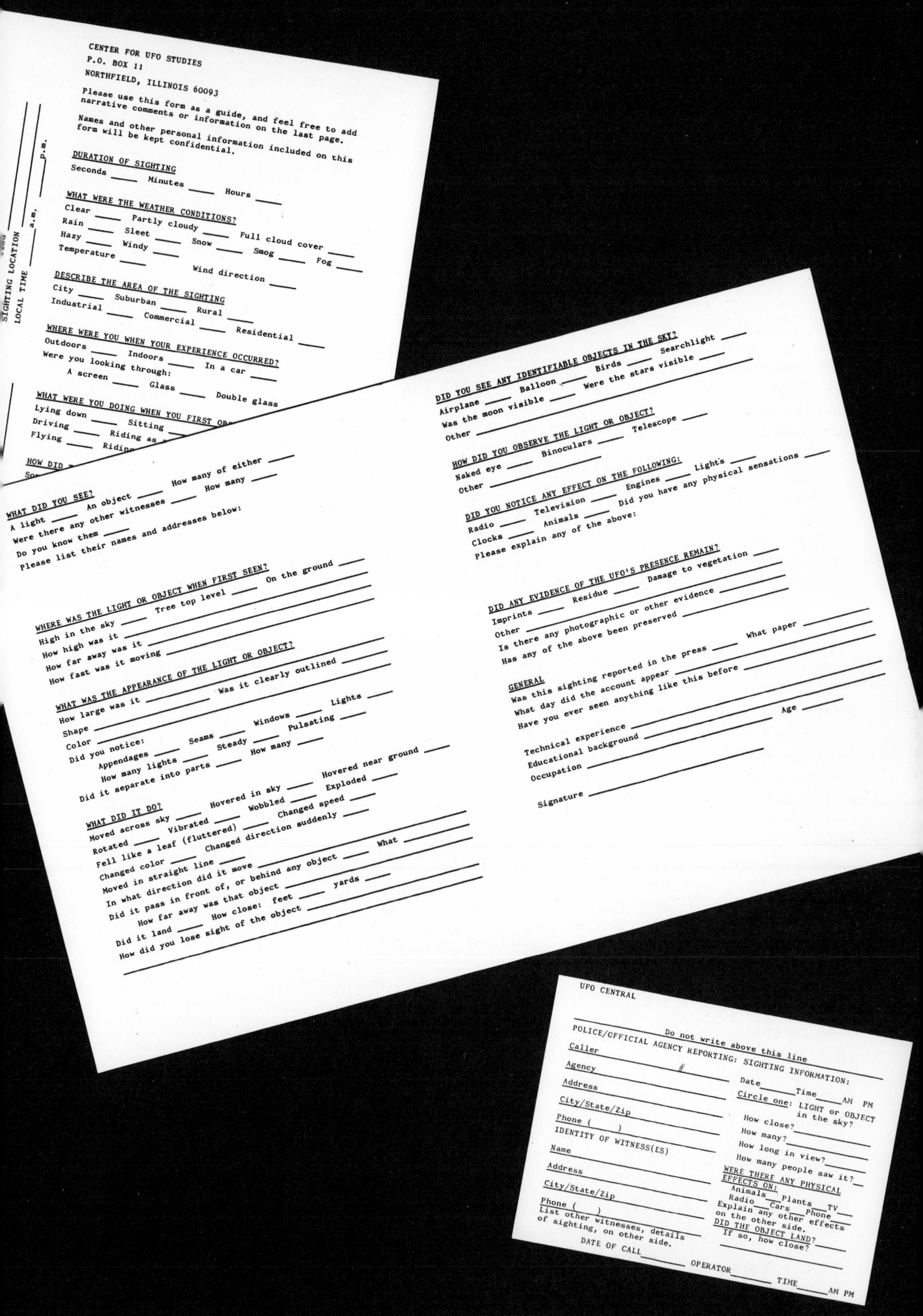

CENTER FOR UFO STUDIES
P.O. BOX 11
NORTHFIELD, ILLINOIS 60093

Please use this form as a guide, and feel free to add narrative comments or information on the last page.

Names and other personal information included on this form will be kept confidential.

SIGHTING LOCATION ____
LOCAL TIME ____ a.m. ____ p.m.

DURATION OF SIGHTING
Seconds ____ Minutes ____ Hours ____

WHAT WERE THE WEATHER CONDITIONS?
Clear ____ Partly cloudy ____ Full cloud cover ____
Rain ____ Sleet ____ Snow ____ Smog ____ Fog ____
Hazy ____ Windy ____
Temperature ____ Wind direction ____

DESCRIBE THE AREA OF THE SIGHTING
City ____ Suburban ____ Rural ____
Industrial ____ Commercial ____ Residential ____

WHERE WERE YOU WHEN YOUR EXPERIENCE OCCURRED?
Outdoors ____ Indoors ____ In a car ____
Were you looking through:
A screen ____ Glass ____ Double glass

WHAT WERE YOU DOING WHEN YOU FIRST OB...
Lying down ____ Sitting ...
Driving ____ Riding as ...
Flying ____ Riding ...

HOW DID ...

WHAT DID YOU SEE?
A light ____ An object ____ How many of either ____
Were there any other witnesses ____ How many ____
Do you know them ____
Please list their names and addresses below:

WHERE WAS THE LIGHT OR OBJECT WHEN FIRST SEEN?
High in the sky ____ Tree top level ____ On the ground ____
How high was it ____
How far away was it ____
How fast was it moving ____

WHAT WAS THE APPEARANCE OF THE LIGHT OR OBJECT?
How large was it ____ Was it clearly outlined ____
Shape ____
Color ____ Windows ____ Lights ____
Did you notice: Seams ____ Pulsating ____
Appendages ____ Steady ____
How many lights ____ How many ____
Did it separate into parts ____

WHAT DID IT DO?
Moved across sky ____ Hovered in sky ____ Hovered near ground ____
Rotated ____ Vibrated ____ Wobbled ____ Exploded ____
Fell like a leaf (fluttered) ____ Changed speed ____
Changed color ____ Changed direction suddenly ____
Moved in straight line ____
In what direction did it move ____
Did it pass in front of, or behind any object ____ What ____
How far away was that object ____
Did it land ____ How close: feet ____ yards ____
How did you lose sight of the object ____

DID YOU SEE ANY IDENTIFIABLE OBJECTS IN THE SKY?
Airplane ____ Balloon ____ Birds ____ Searchlight ____
Was the moon visible ____ Were the stars visible ____
Other ____

HOW DID YOU OBSERVE THE LIGHT OR OBJECT?
Naked eye ____ Binoculars ____ Telescope ____
Other ____

DID YOU NOTICE ANY EFFECT ON THE FOLLOWING:
Radio ____ Television ____ Engines ____ Lights ____
Clocks ____ Animals ____ Did you have any physical sensations ____
Please explain any of the above:

DID ANY EVIDENCE OF THE UFO'S PRESENCE REMAIN?
Imprints ____ Residue ____ Damage to vegetation ____
Other ____
Is there any photographic or other evidence ____
Has any of the above been preserved ____

GENERAL
Was this sighting reported in the press ____ What paper ____
What day did the account appear ____
Have you ever seen anything like this before ____

Technical experience ____ Age ____
Educational background ____
Occupation ____

Signature ____

UFO CENTRAL

Do not write above this line

POLICE/OFFICIAL AGENCY REPORTING: SIGHTING INFORMATION:
Caller ____
Agency ____ # ____ Date ____ Time ____ AM PM
Address ____ Circle one: LIGHT or OBJECT in the sky?
City/State/Zip ____ How close? ____
Phone ( ) ____ How many? ____
IDENTITY OF WITNESS(ES) How long in view? ____
Name ____ How many people saw it? ____
Address ____ WERE THERE ANY PHYSICAL EFFECTS ON:
City/State/Zip ____ Animals ____ Plants ____ Radio ____ Cars ____ Phone ____ TV ____
Phone ( ) ____ Explain any other effects on the other side.
List other witnesses, details of sighting, on other side. DID THE OBJECT LAND? If so, how close? ____
DATE OF CALL ____ OPERATOR ____ TIME ____ AM PM

AFTERWORD

There is no conclusion in this book. It would be presumptuous to draw one.

I have attempted to give a cross-section of the most enduring and the most topical views of scientists on the subject of extraterrestrial life. If I have played down any aspect of the debate, it is the dissenting voice that considers a search for other intelligence a fruitless task.

I believe that there is enough in the debate already to answer that view. We are so ignorant of life-processes in the Universe that any information, even negative, is of value. And, though I must confess myself daunted by the magnitude of the task to seek out extraterrestrial life, there is always that intriguing possibility of success.

There are few expensive and extensive scientific investigations that one can defend on the grounds that they *might* discover something, even though the odds are greatly against them. Yet the search for extraterrestrial life, slender though its chance of success may be, is fired by a deep emotion that is stronger than logic.

As Carl Sagan says: 'The search for life elsewhere is something which runs so deep in human curiosity that there's not a human being anywhere in the world who isn't interested in that question.'

Bernard Oliver's view is unequivocal: 'All past human history may indeed be merely a prelude to an inconceivably exciting future as participants in a galactic culture.'

And, from the more pragmatic point of view, those who decry any scientific or technical possibility are most often proved wrong by subsequent events.

No equipment need be built with the *sole* purpose of looking for extraterrestrial life. There are purely astronomical jobs that the same equipment can do. I foresee that searches for extraterrestrial signals will become an integral part of radio astronomy work at the major radio observatories. This is starting to happen, but it will take a few years yet before the most conservative astronomers come to embrace CETI as a natural extension of their subject. As they do so, however, optical astronomers will work more closely with the radio men to identify likely target stars. And the infra-red astronomers will think more often about the application of their work to discovering planets and technological civilizations.

Eventually, something like the Cyclops array will be built. During the next fifty years, vast data-receiving areas will be needed to keep contact with our first interstellar probes, and possibly to act as communication links with our colonies on the Moon and Mars and with permanent explorer probes keeping station throughout the Solar System. For logistic reasons it is likely we would want Cyclops on Earth, and not in space.

Another reason for fusing CETI research with normal astrophysics is that a message from space will take a lot of looking for. Most of the time we will hear nothing. But during that time we could be learning a lot of astronomy.

The Apollo programme had a visible goal, as would programmes to land on Mars or build a starship. Listening for a radio message that may not even exist is, by contrast, an act of faith—and few legislators would have that sort of faith. So a Cyclops-like array is more likely to be funded to support a tangible endeavour such as starflight or planetary colonization.

Planetary research and starflight are natural adjuncts to the search for extraterrestrial life. In the short run, they may also be the most fruitful. I would be more excited by the discovery of organisms on Mars, which we could actually study first-hand, than the reception of a signal from far away, whose senders we could not see or exchange messages with inside the human lifetime.

If nothing else, the search for extraterrestrial civilizations will tell us more about the origin of solar systems, the history of stars, energetic processes in distant galaxies, the past and likely future of the Universe, the beginnings of life—in short, more about us. Humans are naturally inquisitive and basically narcissistic: such information would satisfy both traits.

Yet what would happen to our curiosity if we began to get information, in forced-feeding fashion, down the funnel of a radio telescope from a civilization elsewhere in space? George Wald, Harvard's Nobel-prize winning biologist, expressed his fears at a Boston University symposium in 1972. He felt that suddenly being told answers that we now need to work for would, in his phrase, 'fold up the human enterprise'. Despite the vast intervening distances, he visualized such an interstellar communication link as tying us like an umbilical cord, or as a dog is leashed to its master.

Carl Sagan countered with typical dry humour, 'Learning that there are a million civilizations out there, all ugly and all smarter, I would call a character-building experience.'

Knowledge that we are not alone, and clearly not special beings, might just engender enough humility in society to prevent us all from becoming too convinced of our own merits. In many people, the medieval—or even prehistoric—urge to regard ourselves as central to the Universe still runs deep and strong.

While aware of the long-duration task that the search for extraterrestrial life sets us, I am still somewhat depressed by the total lack of any direct indications of other life. If the most optimistic predictions are correct, such as Kardashev's Type II and Type III civilizations, one would expect to have found some evidence of advanced life-forms in space.

Perhaps, after all, life is not a major modifying force in the Galaxy.

But if it is, the prophet of astro-engineering activity, Freeman Dyson, has words of warning about what we may find: 'I have the feeling that we're going to discover things that are not so pleasant, particularly since the activities we're likely to discover first are highly technological activities.'

He continues, 'We're more likely to discover first the species in which technology has gone out of control, like a technological cancer spreading through the Galaxy.'

For the moment, though, the exobiologists have to endure the nagging doubt that what they profess to study does not in fact exist.

Iosef Shklovsky summed it up at the Byurakan CETI conference. He said, 'All the natural sciences rely upon observations and experiments.

'We have none of that here.'

WHAT ELSE TO READ

Seminal papers in this field, both popular and technical, are collected in *Interstellar Communication*, edited by A.G.W. Cameron (W.A. Benjamin, New York, 1963); several of these papers have been drawn on in the present book.

Discussion of the whole field by international experts is contained in the proceedings of the Byurakan conference, titled *Communication with Extraterrestrial Intelligence*, edited by Carl Sagan and published in 1973 by the M.I.T. Press. Papers reviewing some of the aspects discussed at the conference were published in the journal *Astronautica Acta* for December 1973.

For an example of space consciousness, get an extraterrestrial perspective via *The Cosmic Connection*, written by the man who is the guru of this field, Carl Sagan (Anchor Press, New York, 1973; Hodder and Stoughton, London, 1974). For my impressions of Sagan, see *New Scientist*, July 4th, 1974.

A well-reasoned case for the validity of U.F.O. reports is contained in *The U.F.O. Experience* by J. Allen Hynek (Henry Regnery, Chicago, 1972; Abelard-Schuman, London, 1972). See also my profile of Hynek in *New Scientist* May 17th, 1973.

Ronald Bracewell has developed his idea of interstellar probes in *The Galactic Club* (Freeman, 1975).

Plus papers on the subject published regularly in the journals cited throughout the text.

INDEX